First of Many Open Academic Letters

ISBN-13: 978-1533410757

ISBN-10: 1533410755

WRITTEN BY PEET SCHUTTE

© KOSMOLOGIESE EN ASTRONOMIESE TEGNIKA

THIS LETTER IS THE ANNOUNCING OF THE BOOK IN SEVEN VOLUMES CALLED

MATTER'S TIME IN SPACE : THE THESIS VOLUME 1-7

ISBN   0-9584410-8-1

TO WHOM IT MAY CONCERN,
An open letter TO SELECTED ACADEMICS ISBN 0-9584410-9-X
is THE ACADEMIC NOTIFYING OF
MATTER'S TIME IN SPACE : THE THESIS ISBN 0-9584410-8-1 Written by PEET SCHUTTE

Dear Professor,

I am Petrus Stephanus Jacobus Schutte going by the name of Peet and who is the author of the above-mentioned book(s). I hope you find your reading of this book presented as an open letter a most fruitful experience. I feel I need to warn you whom are reading this letter that this work contained in this letter strays widely from mainstream science and for that there is a very good reason but I should add that in the least it is thought provoking. I researched the work of a man that is most exceptional and even more prominent in the history of mankind. His role in the gathering of information furthering knowledge accumulating of the human species' efforts stands second to none while most of everyone is not even aware of the full implication of his work. While recognising his work Mainstream science bluntly ignores his work and in that they miss the full vastness of the wide influencing of his work. It is therefore almost absolutely realistic to say that what you are about to read in this open letter sent to you for your attention was never yet printed in the near or the far past although the work has been with us for about four hundred years during which time it went unnoticed. It seems to me that any research predating Newton never came into use or in practise. My investigation of Kepler's work brought about a conclusion that no one yet arrived at concerning the findings of Kepler because no one scrutinised Kepler's formula. Kepler found planets rotating around a centre but Newton saw a circle and added what is mathematically required to indicate such a circle. Newton added a mathematical $4\Pi^2$ to the formula of Kepler and removed the distance symbolising measure that Kepler introduced using **k**. On the other side Newton changed the symbol of **k** by using the symbols G $(m + m_p)$. This is just a longer and probably a more detailed manner of indicating **k** and better defining of **k** but it symbolises precisely to the point what **k** stands for nonetheless.

I wish to draw your attention to the matter of Johannes Kepler's findings that Mainstream science considers as resolved and closed for many a century. Not one of the following principles was yet successfully proven but I believe I have accomplished that goal. I too am well aware that at the first glance you will immediately arrive at the opinion that the theme of the letter has to be considerably below the standard of an intellectual Master such as you must be being in the position you hold and therefore the normal research work you do. Not withstanding I hope that this writing may spark interest even at such a low academic level and grade in scientific sophistication development because I am about to prove that I discovered:

1) The location, the position and the value of **singularity** as a factor forming space-time
2) Finding **space-time** by dissecting Kepler's formula in relation to valuing singularity
3) Finding space-time, **proving space-time** and **aligning space-time** with gravity
4) The **working principals** behind and manifesting **of gravity** as a cosmic occurrence.
5) The **Roche limit**, and explaining the resulting of a law coming about from singularity.
6) The **Lagrangian system**, how and why that becomes the building form of the Universe.
7) The **Titius Bode law** and I show mathematically how gravity comes about from that
8) The **Coanda effect** and the producing of gravity through reproducing space-time.
9) The **sound barrier** by proving it **is gravity** generated **by motion** in space becoming independent where motion creates independence. Breaking the sound barrier is the motion in space duplicating space by crossing over gravity borders. It is $a^3 = kT^2$ where $(k > T^2)$ or $(k > T^2)$

I first started my studies in the field of Cosmology as a spontaneous development of my natural curiosity spawned from childhood interests in the field of cosmology, which I developed even before I went to school. The studies were a reaction (I would imagine) that were part of my personal childhood development in how I was forming a personal concept of a lifelong interest that followed me into my future. At first I conducted all my earlier studying mostly on the basis that inspired me to find out more about what made the Universe tick, with no intention ever on my part to reach a point where I would be writing books on the subject. At first I was investigating cosmology on a part time basis. This went on, on and off, or the best part of twenty odd years (*as* time and *when* time would permit). Then in later life with my health deteriorating I committed myself to more intense investigation and my effort developed onto involving a study using time that is only permitted by a person when that person is involved in such a quest on a full time basis. That quest has now been going on for the last seven years in full devotion and if one includes all the years invested on my part including the twenty odd years before, part time, then the time I have spent in completing my theory when adding all in comes down to almost twenty eight years. This is to say that I did not come to realise what I am about to introduce on a light-hearted conclusion. I mention this because I wish to ensure the reader that he should have no doubt about my most sincere commitment in producing a cosmic theory on matters concerning the start and the working of the Universe during and before the Planck era.

I firstly argued that if there were any way that any one wish to break through the barrier placed on us by normal research it would be in finding a barrier it would be where our vision we use to focus would limit indicate the limit. If we wished on progress in our pursuit of the very first cosmic moment then we have to find and cross the barrier that blocks our view. We have to look deeper and in another direction should the desire driving us be strong enough to commit us to reach into the very birth of the cosmos. We have to rethink the strategy that we use. Max Planck was one of the most brilliant men of all times and even he, notwithstanding all his personal brilliance, accomplished little. There are parts missing in what we have and that which we have at our disposal to use because if there was no such an obvious barrier then the Wise-Men involved in science would by now have found the way to break through the seal that is locking us out of the critical past which will uncover the origin of the Universe's infancy stage. I went about trying to find what everyone since Adam, (meaning all of the rest of mankind and myself) were missing throughout the ages of speculating and interpreting while philosophising about whatever we find inspirational.

The obvious we saw; that was clear. Therefore I had to find a route that would lead into the not so obvious that all of us were missing, notwithstanding the best efforts of the best qualified to accomplish such a breakthrough. My effort involved trying to accommodate that what was in the cosmos available to use by the cosmos in all phases of developing. If I had any hope of finding the answer, such an answer had to be simple because I am not very inclined to unravel what is deemed as complicated. The simplicity had to be locked in what was not yet understood about that which was in the cosmos as it formed part of the process used in forming the cosmos. My realising this brought me to focus not on that which we understand. There is not a lot we actually understand because even gravity is very poorly understood. In fact gravity is so poorly understood that there is not one person alive that can claim the prestige of understanding gravity and among the dead there is even less that can make such a claim.

There are several phenomena that are presented in nature and acknowledged by science but also discounted by science and therefore not presented as accepted science. By admitting that that what we have available to us to use concerning our research of cosmology in an attempt to better our understanding of cosmology, is useless to use, then one realises that not having what there might be makes what we already have useless. It then is useless to use what there is as part of the big picture we are trying to paint because what we use is not really part of the picture. This leads one to believe that the picture of the cosmos Mainstream science is painting, is being painted without painting a full picture.

In my first attempt to understand the full picture of what science was painting I found so many colours missing there was no picture painted that anyone could appreciate. This is what made me decide to go on researching the 'unknown' in the hope it might clarify the 'known' and as the book unfolds you as the reader may agree that I was correct in pursuing the misunderstood and rejected phenomena. Finding the missing phenomena helped me to place the phenomena mentioned above in a theory where the principles also mentioned above form a part of the overall gravity used in binding the Universe. I believe what is in the Universe is not able to be coincidental because of too many influences contributing to what there is - notwithstanding the fact that that is the manner which science uses when they refer to the Bode law. What is in the Universe has a role as it had a role, which is the same role that phenomena has had and in future will have. This is establishing a very new idea about the working relationship between particles and in explaining it by using Kepler's studies. Redefining the work of Kepler's views brings a new Universe to light involving new concepts that are based on old principles but principles in updating man's view about cosmology are very new in that capacity. Through that new vision I was able to come to realise what the reasons might be why Kepler never saw it fitting to include the measure of Π in his formula. I do not suggest his neglect thereof was intentional, nevertheless the formula he devised without using Π proved that there was no need for the inclusion of Π since his figures brought about a correct answer in the final end result leaving a well concluded fitting answer. The numbers he produced brought about a specific space $\mathbf{a}^3$ contained in a circle $\mathbf{T}^2$ at the distance of $\mathbf{k}$ from a defining centre thus the calculations did not require the use of Π to find a meaning. In that Kepler did not see a need to include Π. I would not go as far as declaring with absolute certainty on his behalf that he did it deliberately, however there never arrived such a necessity. It is prudent to agree on whether or not such a need is necessary, because if one is agreeing about such changing not being required a new Universe emerges.

The circle that Kepler discovered came about without ever forcing Π into the frame because it is clear that the circle formation came about as a natural consequence and came spontaneously delivering an equation while he was working. In this book I prove that the reason for adding Π to the rest of Kepler's formula is unnecessary. This unnecessary addition is because when going one step further in the investigation one will find that $\mathbf{k}$ and $\mathbf{a}$ and $\mathbf{T}$ are symbolising the same value with the only difference being that each one represents a different dimension to our six dimensional or six sided Universe we enjoy. In fact I shall show that Π replaces "$\mathbf{a}$" and "$\mathbf{k}$" and "$\mathbf{T}$" and that Π is the true value that should be replacing each factor as to indicate the correct value to the sides nominating Π. We humans work on a numerical base using ten as a basis where we count to nine and re-establish a new decimal numbering line by adding a nought behind the number in value. This is using the numerical basis of ten, which I suspect we took from ancient knowledge about cosmology and not from using our fingers and toes as the earliest calculating processors. In this letter there is unfortunately no room to explain my suspicion but another fact I do prove is that the cosmos uses Π in the cosmic numerical basis as a means to measure and quantify. Therefore in fact the Kepler formula should read instead of $\mathbf{a}^3 = \mathbf{T}^2\,\mathbf{k}$ as it does it must be $\Pi^3 = \Pi^2\,\Pi$ where I shall show that Π represents singularity wherefrom the entire Universe sprang from Π and by forming as $\Pi^3 = \Pi^2\,\Pi$ it is confirming that space is equal to the motion thereof. Kepler's greatest achievement was showing that the cosmos is space –time $\mathbf{a}^3 = \mathbf{T}^2\,\mathbf{k}$ while time is the motion of space in space.

The value of Π is the primeval and most basic of measures applying as an accepted cosmic legal value that the cosmos used exclusively in the very beginning and as it does today. The measure of Π in the Universe, values particle development that brought about all development ever conducted in the Universe. Only after this stage did the rest come including mathematics and went on to freeze spilled singularity into frozen material. Reading this statement may sound suspiciously senseless but as the book unfolds the sensibility will become apparent. The full implication of such a statement will become clear when one dissects different facts coming from studying Kepler. My discovery of this fundamental basis of legal valuing ensured me again that there was no need for someone the likes of Newton to add Π in any form to the work of Kepler because Kepler discovered the ultimate Π in the Universe, the Π giving the Universe form and gravity. The concept of Π that is the only single form of all forms available that can by duplication of Π assemble the value of gravity. When replacing the symbols with Π the facts of the Universe become self-explanatory because the most basic form that forms the cosmos has a definitive and uncompromising value.

But getting this far took me down roads overgrown by ignorance and which I had to uncover myself as if hacking away miles of overgrowth with a machete chopper. All of the disbelief science showed to my work in the past and their refusal to see past Newton made any and all attempts on my part as bad as they could be, strangling and smothering my attempts to announce my uncovering of the newly found insight on my part.

For decades I tried to come to terms with the inability there is in science to explain the cosmos in real terms, when using the science of official reputation. That which there is makes a mockery of science because the undisputable clues left in the cosmos makes what little correct explaining there is available, seem like a comedy of errors, when it is mixed in with all the other near Dark Age errors we still use after so many centuries that provided countless opportunities to revise the old muck. By applying current accepted Astronomy as such the phenomenon found all over the cosmos is still beyond the explaining ability of Mainstream science. This is true and it is a shame because it also is an undeniable fact in spite of the vast knowledge and progress in other forms of science taken in the manner science uses when it approaches cosmology. Cosmology truly lagged behind while the understanding and advancing of physics, mathematics and chemistry as subjects were flourishing. By comparison I saw how little there was available in explaining cosmic phenomenon and how much improvement in understanding the other departments such as chemistry, electronics, medicine etc. could offer as results were coming about from research.

Even where there is a little explaining available in cosmology it turns out that such explaining is confusing to say the least and at best it highlights the manner in which science is applying double standards. For decades photographs were the only progress forthcoming as an addition to improve the meagre field in cosmology and that improvement was artificially stimulating cosmology. By providing a false impression of advancement, everyone missed what and how much was missing…To the connoisseur desperately looking for more than the obvious stirred in with some out-dated misinformation dating back to the Middle Ages, it all seemed as if it was a picture portraying the ridiculous to make the sublime look good. The pictures only proved the opposite of what progress in cosmology will represent. In truth and as such in cosmology the cover up that was hiding the lack of progress about the science of true cosmology was only forthcoming in the improving of electronic optical telescopic advances and spectroscopic progress. There were only photographs carrying beautiful pictures which pleased the less informed except the photographs did not bring progress to cosmology at any intellectual level by promoting insight. The explaining that the photos demanded about the subject had the opposite effect of installing hope because what it did do was underline what lack in any notable progress there truly is in our understanding of cosmology and laws in the cosmos.

While such Hubble telescopic images might seem to be clear as daylight it was more than clear there was little academic value to them. To the person in need of more stimulation than being impressed with pictures of God's marvellous Creation and the sightseeing that always accompanies such pictures, such persons always felt very disappointed. The pictures did give satisfaction to those more easily impressed, but the rest of us seeking knowledge accompanied by understanding the images left us despondent. Although they leave the vast majority in total amazement there are those less impressed about not knowing the 'why' and the 'how' in such amazing pictures. I know the group I fall into may be the greater minority and the majority may only demand the portraying of the images, which is what that easily satisfied group demand. The rest of us rouse with anguish at the lack of information about what is known and what lies behind what those pretty pictures are conveying. Nevertheless there can be no real progress in scientific understanding about the images portrayed by the Hubble telescope, and others, if no-one is able to show the slightest clue of a deeper understanding of what is going on in the Universe. Everyone is almost breathless awaiting the commentating by the most informed which accompanies the magnificent cosmic portraying of God's Creation. When we are portraying the new images, we should also be investigating that what we see that the cosmos is at the moment portraying. The lack of actual believable explanation coming from investigating by means of telescopic imaging should impress one and all, but the impressing must not be based on the colours in the images but the sensible information attached to the image investigated. It is *that* that we wish to see. What we wish to see must at least be accompanied by scientifically backed information, which provides the proven understanding coming from science. When science is employing new explanations with such photos it should also be discarding senseless baggage carried over from the past. Most images contradicted Newton and for saying that, every Academic I ever came across in the past ostracized me. That bothers me little! I know I cannot possibly be the only person absolutely discontented with what Mainstream science accepts as science. Here I refer to the out of date theorising Mainstream science still accepts amongst many others as how they suggest stars and planets are forming. One cannot promote cosmology in honesty and advocate scientific fact whilst dishing up such fairy-tale nonsense to students. Moreover I hold the opinion that amongst Academics in particular there must be many if not most that share my personal serious doubts or have an inclination to share some of them. This I say when considering the overall doubtful picture painted about what there is and what one believes there should be. I just cannot believe those forming the most intellectual group of mankind are unaware of the mismatching facts seen over the broader picture because the contradiction and lack of a plan, makes what there is so very doubt provoking. Newton dismissed the formula Kepler presented as all factors forming motion. That is where the apple-cart derailed.

In honesty we have to realise that we cannot dismiss the whole formula that Kepler produced as being motion. It is so much more than just motion. It is $a^3 = k / T^2$: That is what Kepler brought into civilization for all time to come. He saw space a^3 being in isolation due to the time it uses to move T^2 claiming such space forming independence according to the lines k indicate. Let us look at the factors in more detail before we proceed with the rest of the book.

a^3 symbolises a mathematical interpretation of implicating the three-dimensional space.
T^2 is representing the period or time that Kepler suggested we should use to calculate time that holds the orbiting planet in direct contact with the space in relation to a very specific centre.
k is the space taken from the centre to the end of the line from which the planets must have grown if one accepts the Big Bang growth of particles and the affect of the Hubble constant on all cosmos material. The specific value about the centre is most important because from the specific centre gravity always applies the strongest influence.

One cannot justify Newton's dismissing of Kepler's formula as that all factors only contribute to the motion indicated because that is misleading. We all accept that the true cosmic form *would be* and most probably *is* a sphere. Everyone accepts the universe as a whole as a sphere…but why would the sphere form? What would be the reason why the original form that we devote to the Universe would take on a sphere as a natural form? Apparently our imagination grabs the sphere as form. In all natural events the gravity in that space which stands apart and independent from all other space takes on by cosmic pre-casting the sphere as form of shape … **it is because gravity chooses the smallest space to hold the strongest force**.

I am of the opinion that gravity is about dismissing space to the advance of heat increasing in such a specific and concentrated space using the concentration as measure for the heat as well as the space holding the heat in space. According to Kepler that is what he found to be true. Space a^3 will always be circling space around as T^2 in any position from the centre k. That is what Kepler said when he said $a^3 = T^2 k$. Kepler indicated space a^3 will forever fight for independence and show separate individuality in remaining apart as identifiable cosmic components by means of motion. Every space will cling to independence indicated by k through fighting off the integrating of another coverall unifying unit by applying the motion of T^2! The problem we have to solve in this letter is what will the cosmos use to secure such independence between all particles? What sets space apart from the rest of space? First we have to admit that Kepler was the one that introduced the following.

Kepler gave us the answer to the following but no one ever took notice!
Kepler was the one that discovered **space / time** as $k = a^3/T^2$
Kepler was the one that discovered **singularity** as $k^0 = a^3/T^2 k$

Kepler was the one that discovered **gravity** is holding **space-time** relative by the measure of distancing **k** as $k = a^3/T^2$ and $k^{-1} = T^2/a^3$

Everyone able to read mathematics has to realise that Newton suggested collisions between cosmic structures must eventually come about as gravity erodes the distance separating the cosmic structures multiplied by the product of the mass of both structures from both ends. Newton said the multiplying mass of both structures destroys the distance between the structures by using the eroding force of gravity in the square. The cosmos then must end in a Big Crunch with all material joining together but that joining is not forthcoming at all...and that only indicates how much insufficient understanding there is on offer in cosmology by the educated–to-be-wise-about-these-matters. There is precious little available to explain about their field of cosmology amongst the ranks of Astronomers. So...let's us return to the beginning of cosmology before every one became oh so wise and see what there is to see.

While we are in gravity the manner in which gravity applies in our use of gravity makes us part of the Earth by mass forcing us onto the Earth as a semi unit with all other Earth belongings. Is that which we have truly gravity?

By using mathematics the cosmos spoke to Kepler personally and by the use of mathematics as the medium it provided Kepler with information about the cosmos coming directly from the cosmos.

Much of the proof we use about gravity is part of our perception about gravity gained from the obscure position we have relating to gravity because we experience certain positional fixed conditions about gravity. But are our perceptions about gravity truly correct? We only experience gravity as a factor from the position we have on Earth and _only_ while we are being forced to be _a_ part of the Earth's totality.

The cosmos informed Kepler of another gravity, which the cosmos applies much more widely and is used by nature all over the Universe.

The picture we see coming from the Hubble telescope shows why, in the perfect Universe...but can the Universe be perfect when... we see a radius between the sun and individual planets is not using a regular distance as one would expect of gravity in being a force driven by the mass and in that sense the mass is producing the gravity that always remains even because the mass doesn't alternate. As the mass is never changing on either side, that steady mass has to keep the gravity steady. But in our imperfect understanding of the Universe we find that the radius that should be constant varies considerably proving either that mass somehow adds by measure unnoticed while the structure is in orbit and later allows the same amount of mass to escape undetected; or it's the seasons adding and removing mass at will. This is an absolute contradiction to reality if mass was the factor determining the radius we find between the sun and the planets. This suggests strongly that we'd better be getting very suspicious about the idea of mass contributing to gravity. But in contrast to this, science is unshaken about their confidence in the perfection about facts they use in terms of correctness. It is well known amongst all persons that science only uses dependable and ultra reliable facts coming from sources beyond doubt. Referring to any work done by any scientist will find a remark about science only accepting facts they use to work with. It is accepted overall by all communities that in science those in science use one hundred percent accurate facts or they use no facts at all. If our view was as perfect as science would lead us to believe it then must be the Universe that is imperfect as it otherwise would not behave so mystifyingly. The unshaken confidence science uses has us believing at first consideration that the drawing of gravity should produce an even diameter positioned between the sun and the planets because of ever dependable evenly distributed gravity... but I believe there is a perfect Universe and our understanding carries the doubtful suspicions. Delving deeper uncovers even more contradictions and the level of accuracy contained by our scientific understanding then arouses more suspicion about the correctness of science. Remember Newton changed what the cosmos told Kepler leaving much suspicion as to how far the misdirection takes science. We have to correct the facts we doubt because when correcting the facts they use in science concerning our view about science such correcting brings along a better understanding and then the Universe has to become ever more perfect as one learns to understand the perfect Universe even better. But it does require an open and clear mind and it needs no culture driven preconception that should confirm interpretations about facts surmised even before they are carefully studied. It becomes obvious that Newton never gave careful attention to Kepler's findings because if he did he would have seen what gravity is. Kepler described gravity without using the name later given to the process as 'gravity'. The name gravity was not given by Kepler, where Kepler was the person who first coined the concept of gravity, which was later named by the Englishman that later introduced gravity as another concept but in truth the Englishman only gave it a name. It is important to admit that as far as cosmology is concerned Newton gave the concept the name but *only* the name and not the concept of gravity. Newton's persuasion on matters of gravity as gravity functions between cosmic structures orbiting one another as we find in outer space is inaccurate. What Kepler saw, Newton saw differently and used the opportunity that Kepler left by not giving any name to the process he (Kepler) and Tycho Brahe worked on for two life spans. Newton did seize the opportunity to name what he, Newton, saw but that what Newton saw did not include that which Kepler uncovered. In Kepler's era the name or title was lacking but Kepler established the concept of gravity and the formulation thereof. The concept came from Kepler even before the name gravity was used by Newton to describe in the concept of whatever we today (after Newton) became accustomed to believing what the concept of gravity is about. With the help of Newton everyone since Newton confused Kepler and Newton on the issue of gravity and this confusion even begins with Newton. Gravity might not have been named but became a proven concept and factor after Kepler formulised it, which is before Newton named it.

The concept of gravity that Kepler saw is about the manner in which the structures orbit because there is a space that circles around a centre and this process has kept planets secured, connected and rotating around the sun which is the same concept that is keeping the Universe secure and comes about with a process Newton later named as 'gravity'. This what Kepler saw is not the same as what Newton saw when he saw two objects drawing closer by pulling on each others mass. Then later on Newton named what he thought he saw as the force that Kepler saw but introduced another completely different concept. Kepler saw cyclic formations keeping the Universe together and never approaching each other. Newton ignored what he wished not to see but he changed as he saw fit and what he thought that should be. His experience as a young man drove him to establish a process he formulated as the process that is keeping the Universe together. In that act he caped as much as ignored the work of Kepler, which he also named as the same gravity that he saw as a young man. Why he chose to ignore Kepler's findings on gravity we shall never know but why the world still chooses to ignore Kepler's findings about gravity almost four hundred years after the fact I shall never know. My saying this has literally made Academics ignore me as they would avoid the plague. I am not pretending nor do I exaggerate when I say there were those in Academic institutions that questioned my mental development. Some went as far as seeing me as a joker of sorts and I have correspondence to show evidence to that fact. I know by now while Newtonians are reading this letter I have aroused the tempers of every Academic reading this far, therefore let's see what is being ignored by the Academics which I blame to do just that. .

Kepler said gravity in space is about the area a^3 that would always keep equilibrium with the time T^2 it takes to travel the distance of the full circle position placed by the indicator **k**, therefore adjusting **k** as the need arrives. With **k** shifting in length a^3 will have to readjust and therefore T^2 will find a new relating value each time. This was the finding of Kepler and came after his intense study of orbiting planets.

Before I attempt any investigation into this matter there must be coherence in our agreeing about what gravity is. If you the reader insist that the falling of objects is the only gravity found, your further reading will convince you little. Anything we do decide upon must support the fact that it is gravity that prevents planets from dislodging from the grip the sun has on them. Gravity is not about the sun trying to catch the Earth by attracting the Earth…no, there is so much more to gravity. We must be under no illusions about what gravity is and that being the focus of our discussion and where that gravity is because we have to identify and not confuse the gravity we are looking at. We are now discussing the gravity, which is keeping planets circling around the sun, and stars around specific galactica centre.

In that we do not find one example to use as proof in connection to stars coming tumbling down on galactica centres and crushing into galactica centres. If that is gravity keeping structures in orbit around specific centres we must look at the behaviour of the structures in gravity. We have to find a reason why the planets do not reduce the radius between them as Newton suggested but we must trace the reason why it is gravity, which is keeping them apart because if anything, they are departing as they extend the radius connecting them to the sun. That is gravity because it applies throughout the Universe. The gravity Kepler found is the general gravity that is keeping structures from colliding and in that the principles are avoiding collision or on the other hand avoiding abandoning each other. It is about confirming respect for one another's independence and clearly staying at a predetermined distance while at the same time both are sharing a common space unit. That then must be the defining of gravity we have to study to find the Universal enticing gravity holding the Universe together. By close investigation one will find three factors in urgent need of investigation. There is firstly a centre that draws the object closer. This gravity is clearly a synonym to what Newton saw as gravity. If it were not drawing the object closer the object would not be orbiting around the centre and applying motion. It will draw and absorb all rotating things in its field of gravity.

The fact it does not draw the object into its ranks is because there is another gravity standing alongside this first mentioned gravity. Our recognising the first gravity forces us to accept the presence of another part of gravity. This forces us to recognise the second gravity. When saying this we are not using Newton's cosmic formula concept $F = G\,(M.m)/r^2$ because that can barely be what is out there happening. What Newton saw was falling. If that what Newton saw is the only gravity then whatever Kepler saw including all other parts of everything out there that are spinning around some centre must come closer to one another and connect in collisions. While that is not happening we must start to look past Newton to new grounds we can investigate. We have to go beyond Newton and admit there is more than that what Newton had us believe because it is clear that what Newton had us believe… is not happening. That confirms the presence of the second gravity. The fact proves that everything is departing and not arriving. Even the moon is drifting away from the Earth and this information comes about from the most advanced investigation up to date, including a moon visit and the placing of measuring devices there.

Looking at the gravity intensely we find the roving structure travels in a straight line, which repeats another circle around another centre but because of the influence of a centre keeping the roving structure attached to such a centre the motion allows a circle to form by reforming motion from the original straight line to that of a partial circle. There is a centre, a connecting line travelling between what the two points establishes the specifics of a centre within a circle and the end of the circle. According to Newtonians the centre supposedly draws the rotating object closer. That is half the story.

I suggest we do some deliberation and in deliberating may I remind you THAT NEWTON'S OWN LAWS ARE IMPLIED, and again the planets disobey these laws completely!! In the modern age all evidence points away from contracting and favours eternal expanding.

The latest news confirms that the lot is apparently not coming any closer!

In our manner of considering gravity as a phenomenon we find there are three factors interacting and together the three factors form a balance, which produce and are responsible for a balance between all particles in the Universe. This must be gravity since it seems to be the glue that is holding the Universe intact. We can visually see that as the object moves in a straight line because It counteracts the pulling from the centre by a line that indicates the repositioning each time. In parallel with this it also moves in a circle.

One can only interpret this action as being caused by another line just as strong but counter-directed I motion.? The circle comes into action as a counteraction that is trying to accommodate two opposing directions being evenly strong and from that counterbalancing eventually forms a rotating motion trying to satisfy the direction coming from the straight line in one direction and another straight line counteracting the first straight line. In the motion the straight lines coming from opposing values also forms an immediate circle (though only partly) but the overall complement forms a triangle. This shows a very different picture to that which Newton saw.

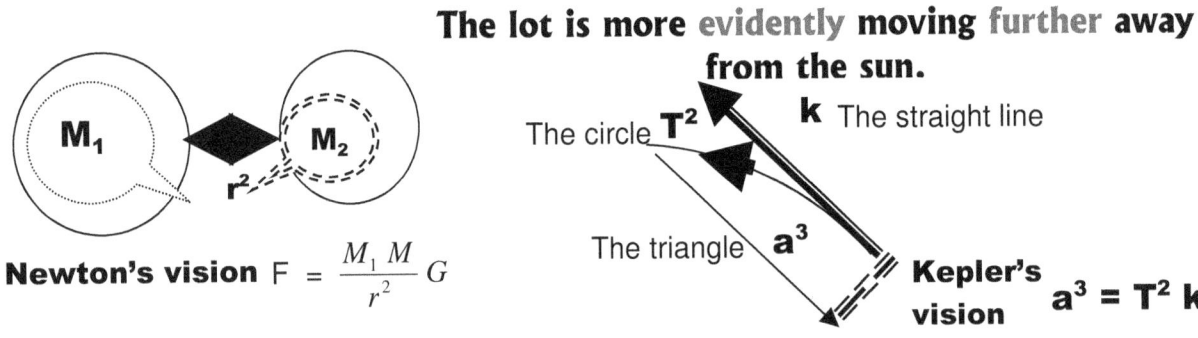

The lot is more evidently moving further away from the sun.

The circle T^2 **k** The straight line

The triangle a^3

Kepler's vision $a^3 = T^2 k$

Newton's vision $F = \dfrac{M_1 M}{r^2} G$

$F = \dfrac{M_1 M}{r^2} G$ This is the suggested formula confirming the behaviour of planets used by Newtonian scholars underlining the argument that contraction is coming about between all cosmic objects. What Newton witnessed, if my memory serves me correctly was an apple falling from a tree where both the apple and the tree were part of the Earth and this did not constitute - or lead to - or come as a result of - a catastrophic cosmic event happening. In the mathematical sense it does not make sense when Newton's argument is taken out and used in outer space. What Newton saw with his falling apple was a mass influencing another mass to reduce the distance as the influencing involved motion that came about. But since the mass in both cases is unchanged and the mass is the factor that is establishing the force that is used by the circle to hold the radius steady and in place, these facts point to a balance that formed bringing about the above-mentioned steadiness. In the view of science however it is the mass that either draws the orbiting objects closer or is keeping them apart. The mass does not change and since that mass of both produces the radius between both, the logic is that there has to be an even and steady radius that develops. The radius has to be equal all the time since the mass never changed throughout the rotation. The radius must be the same from any and all given points that form the rotating circle which must keep the radius equal from every angle…yet we know that Kepler proved this not to be the case even before Newton's naming and changing of Kepler's work came about.

What we see is that there is one factor that is trying to run away being a lesser space within the pulling powers of a larger space (the second factor) trying to capture and control and a referee (the third factor) is seeing to it that the even-handedness is at all times applying in the fight. That gravity which I am familiar with and know is there. In some part but not in all out representing all the gravity there might be because I cannot see the jerking, as much as I do not feel it. That is then most probably

The pulling away of the smaller space. a^3

The double counter-acting referee. T^2

The pulling towards within the larger space **k**

another gravity I can see and which is Kepler's gravity which $a^3=T^2k$ represents. We have a motion of pulling…yes and that is what Newton saw…but then there is another motion of establishing a motion trying to depart, leaving the centre by tearing away from the centre and thirdly there is a motion that sees to it that the balance evolves as rotation. That is what Kepler said when he saw all three factors whereas Newton saw but one of the three. The one space is filling the next space as the space duplicates the position it had in the next moving moment that brings about the next position through motion. This eventually will have confined the next point by using a circle motion, which at first was intended to be a straight line, which is stopped by another straight line. The quest in this book is to find out why the other two factors apply in outer space as only one of the factors comes about on earth under normal applying conditions.

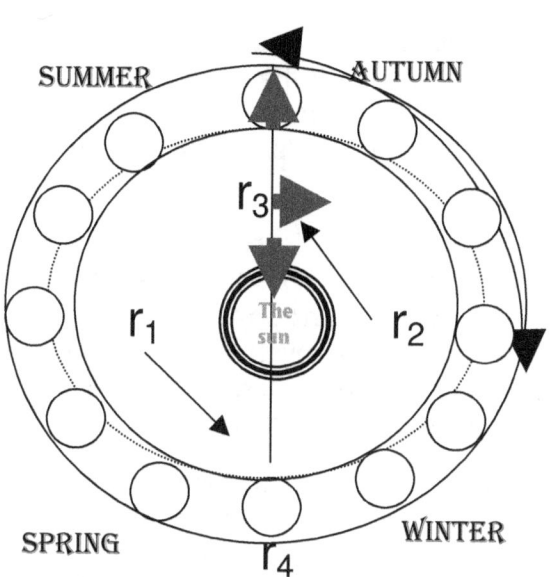

Kepler's investigation indicates to the fact that the orbiting structure is in a motion that is going on where one strength is in a fight with a second strength and the two are pretty much matching in strength because not one of the two is very much winning the dual so no one is winning or losing the fight.

As the two factors are in a motion directional dispute there is obviously one of the two factors or strengths fighting to cut loose from the other one's grip and run off. If there were not such a force trying to escape, the first force would have a quick and decisive victory by reeling in the loser just as Newton predicted. The fleeing object and its matching fighting partner has a third party referee that allows the fight to go in a specific direction as long as there is no decisive victor.

This book, which I produced in the form of an open letter, is on a quest to find the missing two factors and I can declare with some delight and with even more certainty that I found the missing factors. By Newton's introducing gravity as a force with the formula $F = G (M_1.M_2)/r^2$ a precedent was set of gravity being a contracting force forcing distances supposedly to grow smaller. Apply Newton's view to comet behaviour.

Newton insists that the sun has gravity reducing distance between the objects and while lecturers are teaching this during the day, at night they all witness how the comet follows this principle in detail showing Newton as a prophet. No sooner does the final conclusion draw near by orchestrating the final demise of the distance separating the two cosmic components when the opposite changes all concepts taught by institutions of science, the next minute out of the blue with no pre warning of the comet changing its mind, the comet defies all logic in scientific circles that apparently even included defying Newton and his logic. Because at the very point you'd think there is no chance of any return where gravity supposedly should peak because the comet is so close to the sun and due to that fact makes the collision unavoidable…then the comet chooses that very point to dart away into the blackness of outer space, missing the definite collision by miles. By the time the collision is truly unavoidable with the radius between the sun and the comet being as small as it realistically can be the comet starts gaining on the radius distance in spite of Newtonian denial of any possibility that such an event can in fact take place. The radius that should be shrinking further is instead enlarging.

The radius that now begins to stretch proves Newton incorrect and it even depicts Newton as possibly being a fraud. The gravity applied that focussed on the comet reducing the radius between it and the sun was not acting predictably by maintaining the reducing of the distance until collisions come about as Newton insisted on. In our reading the Newton formula in English it says that $F = G (M_1.M_2)/r^2$ which when translated into a verbal dialect that then suggests that a force is committing the material to place a force as a factor involving the forming of a collision. That happens is because of the non-retractable mass that is creating the force, which then makes such a collision imminent. The collision is beyond any attempts of diverting any oncoming objects away from the inevitable possibility of contact between the two objects and ruling out all evading attempts. The destruction possibility as a feature event becomes inevitable. To try and bring anything else into the equation of possibilities is beyond any likeliness preventing all and any rerouting of the approaching objects so that the unavoidable imminent collision that is due to come about in due course will take place between the comet and the sun. That which I just explained is just what Newton suggested. That is not what Kepler said notwithstanding so many arguments with Academics that I had in the past who tried to prove to me that the two visionaries views were equal and the same. Well…it's not the same because when we go onto translate Kepler to the verbal English the letters that come out do not even spell the same words.

Translating Kepler's mathematical expression $a^3 = T^2k$ correctly to the verbal statement in English, Kepler said that there is a **space a^3** which is **equal =** to the motion in the **time duration T^2** thereof between two specific points which is a straight line **k** that holds a relation from a centre to an end where the two ends run from the beginning of **k** to connect at the end of **k.** I might not be the smartest boy on the block but I'm not that stupid either. I know how to translate… and I translate as follows:

a^3 must have a volumetric interpretation because the third dimension is sure evidence of multiple conjunctions of dimensions put together in three sides opposing three sides having the third dimension in place. The fact that any symbol uses a value to the **third power a^3** indicates **space** or a volumetric established and separate unit. Using a cube by three dimensions symbolises a cube, a room, a space to be filled, a unit able to hold other ingredients on the inside when empty or partly filled. It is space because it is volume using the third dimension.

T^2 is an indication of something having a cubic nature other than the square forming motion that is provided by the motion the square indicates, which is where the moving object is representing a third dimensional object that is moving from point to point and it is this point to point that multiplies into the square. The space is moving as a unit from one point to another point and the moving between the points are represented by a flat square or following a flat distance between two points. The cubic space was in one instant in one place and then the second instant in the other and because time can never stand still or become single dimensional (this I am about to prove as the letter unfolds) insisting that time must always support the motion it consists of or time cannot be. It is motion that is taking time, which is motion in the second dimension moving the space in the cube.

k^1 is the symbol used to indicate a straight line between two points with a definite beginning and a specific end position. It is the location where the cube is holding space and where the space was and where the cube in space is going to be in very the next split instant that follows to which will then in multiplying form the square that indicates the time the journey took to move the cube of space from one point where **k** is indicating the location of the space to where the next indicating of **k** will shift the space being the cube pointing at the end of **k,** but since time represents the square and with **k** being the distance that proves that the **k** represents the distance the space representing the cube went to take the time represented by the square through the motion. It is the distance moving space in the cube to complete time in duration in the square of motion; therefore **k** is permitted to be in the single dimension.

There are infinitely more implications in the statement Kepler delivered than what is merely a contribution to motion and only motion as Newton was of the opinion. What is there mathematically not correct in my interpretation of Kepler's manner of translating mathematics to English and why is any changing thereof by Newton or any other person necessary in any way?

We can test any of the following symbolic values in the mathematical expression and also test the principals behind the expression in which Kepler stated them. By such testing we will find that time after time there were never any corrections in the translations required since the translation thereof was never incorrectly presented and in that a case asked for no alterations to secure the correct reporting of the cosmic information being translated.

By taking the formula on face value it can change as follows: $a^3 = T^2\,k$ can become $k = a^3 / T^2$

**Kepler said
$a^3 = T^2k$ but that
could also be
$k = a^3/T^2$**

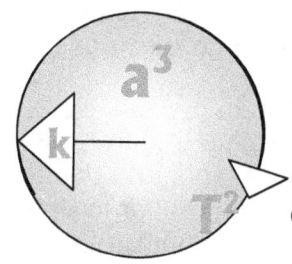

When translating Kepler's mathematical expression into English we can see what Kepler said also read as $k = a^3 / T^2$ where **k** is one point from a centre point that is space a^3 relating to time T^2. From a centre comes space-time. The centre **k** brings space a^3 in ratio to time T^2, which is space / time a^3 / T^2. Reading this correctly cannot bring any dispute...yet it does...and it's been doing it for centuries on end!

With this mathematical reality what then later formed the grounds for any individual to develop any need to change Kepler's translations from the cosmic given to mathematics and then from mathematics to English while the guilty party is renowned for his superior skills in mathematics?

Kepler translated what he found to be the cosmic given to mathematics which we humans are able to interpret from the mathematical expressed to the verbally pronounced and written but Newton still saw a need to change what the cosmos said about how the cosmos is presented and by no one less than by its own interpretation of its self structured composition.

When viewing my interpreting of what Kepler said I might have asked myself countless times what did I not translate correctly from the mathematical expressed to English after encountering a battery of Academic onslaught and resentment on my Newtonian views because after all it is directly diverting strongly from the teachings presented by Mainstream science and the diverting is not coming in a small way.

In truth from my diverting I came across very new ideas I am able to prove. By my translating Kepler's work correctly I came upon answers not yet uncovered by Mainstream Science

Kepler gave the World mathematically translated cosmic answers he received from the cosmos that Kepler uncovered long before Newton, Einstein and others got wise about cosmology...and later the wise came up with old news (old views as far as Kepler expressed their views before they, the wise were born with the purpose of coming to the conclusion that those wise men eventually did) and where the conclusions that the wise concluded brought much surprise to the world with the originality of the later Masters' initiative while Kepler said the same thing ages before...!)

Such is the advantage of recollecting Kepler facts that it does answer many questions, which went unnoticed and therefore not spoken about up to now and some were previously never even thought about.

Newton said a sphere is $a^3 = 4/3 \, \Pi \, r^3$, which is mathematically correct, however

Kepler said the cosmos told him a cosmic sphere is $a^3 = k \, T^2$ There is the two distinct possibilities which Newton saw and which Kepler saw and both are most valid. Between the two concepts there is literally one Universal difference and the two can never be mistaken as promoting the same principles. 'Ever try to answer facts about the Universe in as much as...what brings about the expanding? Kepler said the Universe plus it entire content is expanding centuries before Edwin Hubble realised what he was seeing through his telescope.

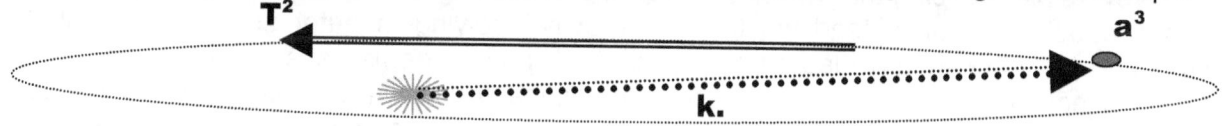

Kepler was the very first person to mathematically introduce **space a^3 centre k** and **time T^2**. Not only did he introduce **space-time a^3 / T^2** but he also placed **space a^3** and **time T^2** in a relevancy long before Einstein did and placed **gravity in space-time a^3 / T^2** even before Newton named gravity. He showed that space **k** is growing in the Universe attend to space-time $a^3 / T^2 = k^1$. Kepler was the person who placed gravity as the ingredient in the universe that determines **space a^3** and **time T^2** and much more. Kepler was the first one that said that gravity comprises of two factors being **k** or linear gravity and **circular gravity or T^2** as gravity keeps space in form while all is staying together.

Although not one Academic has ever openly admitted to me that they as members and part of Mainstream science are more aware than I am of all the facts and doubts I point out to them, such evidence then becomes clear whenever I mention the matter to them I get more than the impression it does not come as a surprise to them and hit them like a brick between the eyes. The lack of surprise and initial doubt they should show at first when they discover the incorrectness of evidence in their theory is a tell-tale sign confirming my suspicions about their evidently knowing all this information all along. They clearly seem very agitated about every detail I show when I bring the mistakes and double talk to their attention in the hope that they may confirm my doubts.

Never is there a whisper of a surprise or a hint of a suggestion that would initiate an argument carried on by the bewilderment or the astonishing surprise they should feel confirming my arguments because there is a mild complacency in their voices. My jumping them total unexpectedly about matters they never contemplated in the least leaves them unturned. The rush in blood pressure that should be a factor on their part and part of the instant where total surprise will bring about some confusing thoughts that will inspire the unleashing of an argument in defending their holy grail should at least carry a surprise in an attempt to save what they believe as being the Gospel in science and with that defending their honour. They lack embarrassment, which they should have in their disputing of my claim as they fight off my allegations with a countering of denial claiming foul on my part as they are in shock when finding out about any doubts.

A lack of true emotion on their part is a telling sign that they also may have some serious thoughts on the quiet about any inclination presenting a flawed view about what they always thought they knew to be true. There is only that eerie dismissing of the seriousness and the lack they show in excitement that would deny or support my credibility as I present my findings. If they know about the inconsequential facts in science why is it not generally acknowledged and pronounced as a matter of fact? Why is there the covering up and hiding facts that we associate with some professional criminals such as politicians. The fact that Academics are aware of this evidence in general terms about the misinformation and doubting evidence about Newton's cosmic vision but moreover underlying this is their total denial of knowing about it and that is what is so seriously unforgivable. The fact that all Academics are aware of my evidence even before my presenting them with such evidence is beyond doubt. If that is the case then why are they forever trying to kill my viewpoint and forever try to silence me where I am only the messenger because I bring the solution and the answer? Please note that the answer and the solution are unbelievably simple and unsophisticated. It lacks all the splendour and grandeur expected by all Academics concerned. It is because it is so simple that it went amiss for four hundred years. It is because it is so simple that it misses the grandeur that will entice them. Instead every academic accuses me of not understanding Newton while they can't show me what part it is that I can't understand and I on the other hand can't see what there is not to understand..

Newton said that it is the reducing of the distance between the objects that would bring about the un-reversible reducing that will end in a total demolishing of the radius that is between the cosmos structures, but instead we find the gravity applying in outer space is one of the instances where gravity provides an orbit circle that gravity seems never to completed as the orbiting objects follow from closing any circle that is leading into a following circle up to where the circle is completed in cyclic precision. That is not the gravity that Newton identified although Newton admitted that there is a presence of a centre forming a point in the middle between the two objects. He was unable to know what caused or even the presence of the Coanda principle, which forms so critical a part of my theory. The formula concerning cosmic balanced gravity however leaves no room for the admitting of such a point and by not leaving a possible inclusion of such a point in his formula Newton did by such gesture in principle repeal his admission of such a centre. This had me cast doubt on what is taught at institutions of learning. It motivated me to venture back to an era before Newton came to influence science. I came to acknowledge Kepler as I came to understand Kepler but it involves much more reading into what Kepler said by finding what Kepler did not say in the way that he did say what he said than it is reading about what Kepler said as it is written as he directly stated what he said. Again I must stress this point: when I refer to what Kepler said it most likely means reading into and is being a part of the part that what he did not say when he was saying what he said but I accept that he meant to say what I am reading as part of what he did not say but meant to say. I have to read more with my mind than with my eyes. This comes as a result of interpreting Mathematics to the verbally expressed. I had to learn to read with my mind and not my eyes and I found that that is the manner in which one has to approach cosmology. From the first time I discovered what manner one should use if one wished to read into Kepler's findings I saw Kepler was all about uncovering the unknown. Realising that, the conclusions I drew by reading in such a way cemented my better understanding of Kepler's work, which then helped me improve my insight into Kepler's work as it increased my understanding about cosmology several fold. This helped me to realise what implications were to be found underneath Kepler's discoveries. From my realising what approach I should use, it helped me to improve my cosmic realising by using the method of reading Kepler and from that I could come to appreciate what Kepler introduced.

Only then did it bring insight and proof to me as a student of Kepler and this proof I found by dissecting what Kepler *did not* say instead of what he *did* say, which I now present to you with this letter, you being a superior intellectual person. Kepler said $a^3 = T^2 k$ and that correctly translates to a mathematical expression $k^0 = a^3 / T^2 k$ which in the verbal statement in English translates that Kepler said that there is a **space a^3** which is **equal =** to the motion in **the time duration T^2** thereof between two specific points which holds a relation onto a centre k^0 where from there forms **a straight line k** that is centred on the spot where space begins from k^0 **that produces k** as well as producing the circle therefore that spot $k^0 = a^3 / T^2 k$ has hold k^0 at a value of having the least space. The line **k** is centred onto a spot where space begins specifically at k^0. This point not only produces the line k^0 but represents also the space that forms the eventual circle T^2. Therefore from the centre holding k^0, k^0 leads to **k** that forms the roving space a^3, which is rotating at a distance **k** where T^2 forms the outer limit of k^0. Mathematically $a^3 = T^2 k$ will be $k^0 = a^3 / (T^2 k)$ because $k^0 = 1$. But $k^0 = 1$ also present the single dimension where all factors are a product of one. If one can locate k^0 one will find singularity. That is where gravity is because gravity is strongest where space is least. Then that suggests that gravity is strongest at k^0 because space is least. That is gravity because that is what keeps the orbiters in orbit but also that is what Newton completely missed when he changed Kepler's work. Newton failed to recognise gravity as the only ingredient in Kepler's formula. He admitted he missed this because he admitted he did not know what gravity is while Kepler explicitly showed what gravity is. Gravity is what keeps the orbiters orbiting. $k = a^3 / T^2$ is **distance^1 = space ^{3}/ time^2** forming from a pivoting centre k^0. That is a cycle and moreover it is a cycle formed **by space/time**. What Kepler said is that space is a^3 **in motion T^2 k.**

That says **space^3 (a^3/)** relates directly to **time^2** that uses the symbol T^2. This is also what I refer to when I say one has to read what Kepler did *not* say when one wishes to see what he *meant* to say. Kepler introduced space^3 –time^2 long before Einstein's date of birth appeared on any calendar although Einstein is credited with the formulating of the concept of space-time and giving it a name. Going even further Kepler stated that the space a^3 is on the move T^2 around in a circle at a distance **k.** That is what that comet we are discussing is doing. The space^3 (Comet) is circling the sun using a radius **k** to establish the cyclic time^2 as a period of continuous motion and continuous motion is gravity. That reads much more correctly and closer to the truth than what Newton predicted what according to him (Newton) was happening in space. Remember in this statement I am separating cosmic principles applying from the way that gravitational principles apply on Earth. I distinguish that which is the rule in the cosmos from what we find ourselves trapped in on Earth. The two just don't mix. I am removing cosmic physics from normally accepted physics because the gravity concerned is not the same.

The proof I bring is real however simple it may seem. It has none of the mind-blowing complexities normally associated in the presenting of investigative analyses of Astronomy. I realise the information in this book carries the arguments in a childlike manner which are very simple to follow, and for that in the past I have been blamed over and over again as being unprofessional. In my answer to that I can only reply by using another question: Are only professionals adequately equipped with minds that make them (the professionals) the only ones able to think? We being part of the human race are all thinkers. Everyone as a human being can think. Every person on Earth is a thinking thinker that uses his brainpower by exploring thoughts mainly and normally to his or her personal benefit. It is what we think about that produces the results of our efforts by which we accomplish what ever we are thinking about. I have met professional Academics that I found foolish as much as there are other cases where the so-called amateurs can credit themselves with much wisdom and insight. Albert Einstein as a patent clerk was that much but to name one. Please understand that I do not compare my achievements or myself in any way, shape or form with the likes of a Master such as Einstein although I speak my mind when not being totally in agreement with some of his or other views. My unsophisticated retracing of Mainstream physics concerning the Big Bang in detail helps to reinvestigate established principles and moreover investigate proof in the light of modern evidence.

In principle I distinguish between Kepler and Newton in that Newton is one hundred percent correct concerning gravity on Earth but as far as outer space forms gravity the conclusions of Kepler and Newton do not match and they had totally different ideas about what they saw in gravity. I am in disagreement with some basic principles that science acknowledges and I divert strongly from all accepted roads Mainstream physics follow. By my doing that those who are considered and accepted as self-proclaimed members of Mainstream Physics have categorised my views in the past as incoherent. That I do not accept. I admit that my line of thought is extraordinary and controversial but only to Mainstream science and not to the standards laid down by nature. Since the concepts I follow start at the beginning, and I take Kepler at the point where modern cosmology began and in that mindset I re-evaluate Kepler's work. I start by tracing a new approach as to what I see Kepler found. The main condition of my investigation is to establish a divorce between what Kepler said and what Newton thought to add to what Kepler said. It is this divorce I create that Mainstream science finds repugnant or even in some persons' opinion repulsive.

I believe the repugnancy does not come from or is not manifested in any part of my work to the letter as such, but rather what my work suggests and who is doing the suggesting. To my view in cosmology such adding to Kepler by Newton was unnecessary and it diverts Kepler's work away from cosmology. But as the generations moved on Newton became religiosity in the mind of science wherever science was taught. To students there is little or no choice in the matter since the only choice left to them is one of understanding by forcefully accepting or die an academic death since Newton is academically accepted without asking questions or raising an opinion. For the second choice, the less accepting students are greeted with a Dear John good-bye letter sending them off into the unknown sunset that such a future outside physics will bring them. That is brain washing.

From studying Kepler I saw that we have to gauge what we find in the Universe. What we find is not that what we realise with our eyes but that what we observe by using our minds to translate from visions coming from our eyes to our minds. We have to test what we are seeing with our minds and read into that which we only can see in the mindset as we see what we can observe by looking at what we cannot see.   This I first found to be true about Kepler's work and when I started projecting this method of observing what the Universe is, as it scattered most previous perceptions I found that using the new method brought along answers so fast I could sometimes hardly keep up with the interpreting thereof. But as is the case with Kepler so is the case with the entire study of cosmology: One should see what there is about the cosmos which is unseen to us and then we may find so much more in the cosmos unseen to us representing that which we cannot see and that which we cannot read because we have to learn to read what is not written in light. Armed with this realising I then proceed from that point by further arguing and debating the full implication of Kepler's contribution. Kepler placed cosmic structures in relevance to one another and so does the Big Bang Theory. The backbone of the Big Bang is that relevancies apply in dynamics and such dynamics are placing all structures without any reservations independent from each other. As the Big Bang progresses all inside the Universe is in the same Universe that will always be the same, however the relations that the elements comply to bring across new relevancies with new positions to fill.   The father of the Big Bang concept is a person by the name of Father LE MAÎTRE, GEORGE ÉDOUARD (1894-1966) who was a Belgian priest and cosmologist. He was the first person to embrace the fact that the universe expanded from an infant stage. His model of an expanding Universe (1927) was superior to that of W. de Sitter in that it took into account mass, gravitation and the curvature of space. Similar models were proposed in the early 1920s by the Russian mathematician Alexander Alexandrovich Friedmann (1888-1925) but Friedman compiled various such possibilities.   Lemaître argued further (1931) that the quantum theory supported an origin in the explosion of a 'primeval atom' or 'cosmic egg' into which was originally concentrated all mass and energy.   As modified by A.S. Eddington, Lemaître's model provided the springboard for G. Gamow's Big Bang theory. In the wider picture of science in general a lot changed to just allow such turnabout in thought since the day of Isaac Newton. From Newton's attraction and contraction many things came into place that allowed change in the most hardened minds. Accepting facts about the Big Bang concept is quite radical. By promoting expansion the Big Bang theory contradicts gravity and our accepting of the Big Bang has to change all other concepts. By accepting the Big Bang other changes are also involved.

KEPLER, JOHANNES (1571-1630)
The German mathematician and astronomer KEPLER, JOHANNES   (1571-1630) became Tycho Brahe's assistant in Prague in 1600 A. D. where he undertook to complete the tables of planetary motion Tycho had begun. Kepler first calculated the orbit of Mars. He spent much time trying to reconcile Tycho's accurate observations of the planet with a circular orbit, but concluded (in Astronomia nova, published in 1609) that Mars moved instead in an elliptical orbit.   Thus, he established the first of his laws of planetary motion. A theory that the Sun controlled the planets by a magnetic force led him to the second and third of his laws, which were published as part of his treatise on theoretical astronomy, Epitome astronomiae Coernicanae (1618-21). The Rudolphine Tables (named after Tycho's patron, the Holy Roman Emperor Rudolph II) of planetary motion appeared in 1627 and were still in use in the 18th century. Kepler also wrote De Stella nova, on the supernova of 1604 and Diptirce on optics and the theory of the telescope. The overall view followed in this book **an open letter To Selected Academics ISBN 0-9584410-9-X** places the true significance of his work in true contents. In KEPLER'S EQUATION is the equation that relates the eccentric anomaly of a body in an elliptical orbit to its mean anomaly.

The equation is $E - e \sin E = M.$, where E is the eccentric anomaly, M the mean anomaly, and e the eccentricity of the orbit. It is important as one of the mathematical relations enabling the position of a planet about the Sun, or a satellite about is planet, to be calculated from the orbital elements for any time. However this only relates to the solar system, and KEPLER'S LAWS only apply in the contents of the solar system. The three laws governing the orbital motions of the planets, discovered by J. Kepler is as follows: The first law states that the orbit of a planet is an ellipse with the Sun at one focus of the ellipse. The second law states that the radius vector joining planets to the Sun sweeps out equal areas in equal times. The third law states that the square of the orbital period of each planet in years is proportional to the cube of the semi major axis of the planet's orbit. The first law gives the shape of the planet's orbit; the second describes how the planet must continuously vary its speed as it follows its orbit, moving fastest at perihelion and slowest at aphelion. The third law gives the relationship between the planets' average distances from the Sun and their periods of revolution.

Instead of studying the true value and contribution of to Kepler's laws an Englishman going by the name of I. Newton placed his own interpretation to Kepler's laws, and in doing this, he wilfully destroyed the principle working of the Creation. Saying this I hear the alarming hooters announce Newtonian dismay. In the past my experience was that all the revered Academics lost their appetite for any further investigation of my work. That is sad as much as it is regrettable. Through Newton's tunnel vision, he applied his own misinterpretations to the correct presumptions of Kepler and through the Newtonian tunnel vision Academics did not move an inch away from repeating the same procedure. In the past it was this that had Academics shying away from me because at the point where I criticise the Newtonian viewpoint and where I declare my suspicions concerning the accuracy and correctness in theorising their way of reasoning which I might add in the past caused the Academic Newtonian wizards mad as bats and then that got me and my work rejected to a point where the applecart lost its wheels on every occasion. It is where Academics read my remarks and what brings (seemingly in an instant) wrath to Academics.

I say this because I realise that reading my remarks or hearing me remarking about this notion brought much resentment on their part and if the reader at the present moment is a Newtonian, boiling his/her blood. It is blood boiling because I believe they see my remarks as belittling that which they feel they have accomplished. This is not the case but still my remarks have the same effect on the Academic as pouring icy cold water down the back of his shirt. I mention this because I know it has happened many times before and if possible I wish to avoid this response. Therefore I ask you kindly to please be warned about the negativity you must feel towards me where you are the Newtonian and I am not. Before you lose interest in reading this letter any further please allow me to finish. In the past Academics thought me to be presumptuous and that normally became the point where all the Academics find their interest vanishes. That should not be because if Newton's work is as utterly accurate as those with faith in his work believe it is, then every aspect about Newton should stand above any and all reprimanding or any form of doubt causing a notion to reprimand. The testing of Newton's work should withstand all testing notwithstanding the person or the prominence of such a person's social or academic standing in the Academic society or even the prominence that such testing will deliver. From what I see about Kepler's work it is a flow of circumstances that lead to Academics neglecting Kepler's work and the realising of the theory I suggest is not forthcoming due to my personal brilliance.

I do not consider myself to be the brilliant in any way as to be the one that can remove the verbal splinter from the eye of the Academic. Yet…if there is a splinter what else should I then do…Newton reduced the implication that Kepler findings hold by introducing to the law of gravitation. He then went about and changed it to three laws of motion. It is clear that while he formulated the laws on motion he missed the way Kepler introduced gravity as space a^3 coming about through motion T^2 and that gravity is space a^3 within space k within motion T^2. Newton also missed the fact that gravity is at its strongest where motion and space cease to be. This is most important to recognise about gravity in one of the two forms it has. I. Newton generalized Kepler's first law, verified the second law, and showed that the third law should be amended to the form; $4 \pi^2 a^3 / T^2 = G (m + m_p)$. In this, the value of "T" and "a" are the period of revolution and semi major axis of the orbit of a planet of mass m_p about the Sun of mass m, and G is the gravitational constant.

It should be clear to any person investigating Johannes Kepler and his work that Isaac Newton hijacked Kepler's work and any time there is the slightest referring to Kepler about the research Tycho Brahe and Johannes Kepler did such referring to Kepler always lead to and always include the mentioning of Isaac Newton changing the work of Johannes Kepler. It is as if the World never could acknowledge Johannes Kepler because the work of Johannes Kepler would be completely wrong and misleading if it were not for the intervention of Isaac Newton saving the skin of the less admirable Johannes Kepler. This comes in the midst of every one realising that Kepler used the information he received directly from the cosmos. I do stress this on many occasions throughout the letter because the embarrassing part is that Newton changed the work of The Universe and not of the man called Kepler. Should you reading the letter entertain the opinion of Newton and feel any urge to defend Newton you should ask the question as to who is standing corrected, is it Kepler or is it the cosmos that gave Kepler the information he concluded? The cosmos supplied all the information by using mathematics, which Kepler then had to translate. But Newton destroyed the accuracy by altering what the cosmos said and directly by adding to that what he (Newton by name) thought that the cosmos left out. This set a precedent by Newton in cosmology and also set a trend, which was retained in all future cosmological development and it lasted in cosmology for three hundred and fifty years. In this book you are reading I am about to show that such practise should no longer be accepted in cosmology. In the process the world of Mathematics developed and the world of cosmology stood still for almost four hundred years. Faculties contributing to cosmology and feeding off cosmology improved as much as they developed, but when cosmologists see the Roche limit in action in the lens of the Hubble telescope and refer to the event as "stars blowing bubbles" being the ultimate response coming from those persons who are supposedly the Masters of cosmology affairs, then the truth of what I just said comes down on you like a ton of bricks. Everyone having any remote interest in cosmology will find they are being very disillusioned by such "official" testimony about the evidence the Ultra Wise report on. This book is about showing how great Johannes Kepler was and how enormous his work was. It will show he preceded all ideas of everyone that came later and officially introduced the novelty of such ideas. Back during the time Kepler was introducing his work the stature and the magnitude of his work was beyond any person's understanding (including Isaac Newton) and this prevailed for most of half a millennium.

I do not say I am the brilliant one to uncover Kepler in the face of everyone failing that came before me, but as I am not a Newtonian such bias was not part of my repertoire and denying me the fortune of being a Newtonian added to my fortune of realising Kepler. Yet as you will notice, the work I contribute is much below the sophisticated norm of modern investigative research and the levels that modern research accomplishment demands to better the effort of the understanding ability in the splendour that investigative research work should deliver in view of our modern times. It is only pure neglect in science circles that moved science past Kepler. Not seeing and therefore not investigating through almost half a millennium has paved a road past the inferior levels that the researching of Kepler's work holds because it was rocket science four centuries ago but the brilliance of it has faded since then. My contribution holds no astonishing flair that may add to science in general. Only failure to notice what I see on the part of those truly brilliant can explain my being able to present my contribution about my work in investigating Kepler. Only by their passing such degrading levels of the Academic establishment in the past and the present can bring the blame for such an obvious discrepancy because any involvement in the work at such an inferior level as that which I bring cannot interest and excite a salted Academic and when thinking about it, the idea is totally unthinkable.

This letter, although it is on this inferior level is about correcting this tendency and has in mind the effort to put in writing what would place Kepler in the greatness and glory he deserves. As I already said, if Kepler was wrong then the cosmos was wrong about facts and applying relevancies and tendencies in the cosmos. I yet again wish to reiterate we should never for one moment forget that Kepler received his information directly from studying the cosmos so how could the cosmos stand corrected? In spite of all the brilliance attributed to Newton nonetheless if Newton had the mind to change Kepler's work and my saying this includes all persons agreeing with such changing by Newton of the work of Kepler those persons admit that he or she or Newton never took any time to really and truly investigate what the cosmos told Kepler. From my reading into the work of Kepler I prove gravity, the Titius Bode law, singularity, space-time, space-time relevancy, the Lagrangian system, the Coanda effect and the Roche principle, the sound barrier, the principle behind the Black hole. The precondition for my ability in doing so is that I have to remove Newton's opinion about Kepler's work from Kepler's work. Whenever cosmology comes into question and all the phenomena, which I mentioned just now remains unexplained and by that token alone it shows to what degree did cosmology remain undeveloped. Whenever there is any mention of Newton, Kepler is never mentioned. But the reverse is always applying. Mainstream physics holds the opinion that Kepler may only have an opinion if Newton can change the opinion. Kepler gave space-time, gave gravity, gave singularity, gave the Plank theory, gave the theory on relativity but no one ever found Kepler's work deserving enough to launch any investigation such as I did. I belabour this because of what revulsion my rejection of Newton unleashed. That is one barrier much unnecessary but it has been an insurmountable barrier this far.

NEWTON, ISAAC (1642-1727) and NEWTON'S LAWS OF MOTION
An English physicist and mathematician who developed his principal theories about gravitation, optics and mathematics between 1665 and 1666. In 1668, he made the first working reflecting telescope. Most of his work remained unpublished for long periods, partly because of criticisms by c. Huygens and the English scientist Robert Hooke (1635-1703) of his early work on the corpuscular theory of light. However, in 1684 E. Halley persuaded him to organize his work on the celestial mechanics of the Solar System, which was published as the Principia. Newton's other major work, Opticks, was not published until 1704. It contains his corpuscular theory of light, and the theory of the telescope. His greatest mathematical achievement was his invention of calculus, independently of the German mathematician Gottfried Wilhelm Leibniz (1646-1716). His profound influence on physics and astronomy is reflected in the phrase 'Newtonian revolution'. Three laws published in 1687 by I. Newton concerning the motion of bodies.

1. A body continues in a state of uniform rest of motion unless acted upon by an external force.

2. The acceleration produced when a force acts is directly proportional to the force and takes place in the direction in which the force acts.

3. To every action there is an equal and opposite reaction.

4. However there is one more law on motion that went undetected by Newton...This book is not about trying to disprove Newton...it is about adding too science more than there now is available without removing any that science already accumulated.

In this book I use Kepler's formula to either prove or to disprove the following accepted principals in cosmology and if any person in the past gave only the slightest attention to Kepler's work, many statements would have come much sooner delivered by someone else or may never have come at all. By applying Kepler's formula correctly in this letter I can either agree with or in other cases deny the following principles.

It began with NICOLAUS COPERNICUS who changed the status quo. COPERNICUS, NICOLAUS (1473-1543) was, according to the Anglo Americans, a Polish churchman and astronomer although this is just more politically inspired propaganda because his parents were both German (in Polish, Mikolaj Kopernigk). While he was completing his studies, he had realized that the Earth revolves around the Sun and not vice versa. Such a view was in that time, held to be heretical. As I pointed out in the first few articles, the Church regarded the geocentric world-view of Ptolemy as consistent with its doctrines. Copernicus set down his basic ideas around 1510 in the Commentariolus, which he circulated anonymously, because of the Islam link. In 1512-- 29 he conducted his study and concluded the observations that he needed to support his theory, while carrying out ecclesiastic and local administrative duties. In this time, he had to defend his mother in court on charges of witchcraft. In 1539, the Austrian astronomer and mathematician Georg Joachim von Lauchen (1514-74), known as Rheticus, became a pupil of Copernicus and began to spread his ideas. The published work was openly spread as the Copernican system, in spite of the life-threatening dangers connected with such a "crime", in 1543 in the book De revolutionibus orbium coelestium. However, the reality of a heliocentric Solar System was only commonly accepted, after the work of Galileo and J. Kepler. The ideas introduced developed along and proved to be correct until such a time it met a solid wall with the investigation of Max Planck.

PLANCK CONSTANT
(Symbol h) A constant that relates the energy of a photon to its frequency. It has the value 6.62076×10^{-34} Js. It is named after the German physicist Max Karl Ernst Ludwig Planck (1858 – 1947). PLANCK ERA. In the Big Bang theory, the fleeting period between the Big Bang itself and the so-called Planck time when the Universe was 10^{-43} s old and the temperature were 10^{34}K. In this period, quantum gravitational effects are thought to have

dominated. Theoretical understanding of this phase is virtually non-existent. It is named after Max Planck (1858-1947). PLANCK'S LAW

A mathematical description of the energy radiated at different wavelengths by a black body: $E = hf$, where E is the energy of a photon and f its frequency. It was formulated in 1900 by Max Planck (1858-1947), who realized that energy is radiated in discrete packets, which he called quanta, and it formed the basis of quantum theory. The quantum of light is a photon, the energy of which depends on its wavelength.

One rule which is well established and which Mainstream science all agree is one aspect which forms the very principle that holds the theory about the cosmic start together under the covering of a verbal blanket. All agree that it all started with singularity but I manage to go one step further where I prove that it is also where it ends, as singularity reunites space-time, which is where Creation split in the very beginning.

Singularity is as follows: Singularity: a mathematical point at which certain physical quantities reach infinite values, for example, according to the general relativity, the curvature of space-time becomes infinite in a black hole. In the big bang theory the universe was born from singularity in which the density and temperature of matter were infinite. From singularity flows space-time.

Space-time is as follows: Space-time is a four dimensional position of the universe where the position of an object is specified by three coordinates in space and one position in time. According to the theory of special relativity there is no absolute time, which can be measured independently of the observer, so events that are simultaneous as seen from one observer occur at different times when seen from a different place. Time must therefore be measured in a relative manner as are positions in three-dimensional Euclidean space, and this is achieved through the concept of space-time. The trajectory of an object in space-time is called world line. General relativity relates to curvature of space-time to the positions and motions of particles of matter.

SPECIAL THEORY ON RELATIVITY
A theory proposed by A. Einstein in 1905, based on the proposition that the speed of light in a vacuum is constant throughout the Universe, and is independent of the motion of the observer and the emitting body. A consequence of this proposition is that three things happen as an object's velocity approaches the speed of light: its mass goes up, its length shortens in the direction of motion, and time slows down. Hence, according to special relativity, no object can ever reach the speed of light because its mass would then become infinite, its length would become zero, and time would stand still. In addition, Einstein concluded that the mass of a body is a measure of its energy content, according to the famous equation $E = MC^2$, where c is the speed of light. This equation describes the conversion of mass into energy in nuclear reactions within stars.

GRAVITATIONAL COLLAPSE
The collapse of a body that is unable to support itself against its own gravity. Gaseous bodies undergo such collapse if they are not hot enough for their gas pressure to balance gravity. This can happen in the early stages of star formation, or when nuclear burning ceases in a star's core. The time taken for such collapse decreases rapidly with increasing density, varying from about 100 000 years for the birth of a new star to less than a second for the formation of a neutron star. Star clusters may undergo a similar collapse if the random motions of their constituent stars are insufficient to offset gravitational effects, either during their formation or at an advanced stage of their evolution.

GRAVITON
A hypothetical particle or quantum of gravitational energy, predicted by the general theory of relativity. Gravitons have not been observed but are predicted to travel at the speed of light and to have zero rest mass and charge. A graviton is the gravitational equivalent of a photon. It is this anti-photon-being-a-graviton by just merely swapping direction and all is proved that I find not very indigestible in modern science. One of the main issues that I wish to protest by

my writing of this is my argument that if the Universe can be compressed back to the size it had at the point of 10^{-38} seconds after the Big Bang the daily outdoor temperatures of 10^{27} K will also come about once more. The expansion was the result of compressed space, which then formed heat and in turn resulted in finding a Universe with the insufficient space prevailing at the time wherein space growth was the converting of such heat into space. If the Universe was in a vacuum as big as being available now then what was the temperature of the vacuum while it was empty before material filled it later. Then I presume the vacuum was there present as it is now in this present day. If the Universe then employed the space of say one atom, the impression comes through that from edge to edge and from Universal border to border the space occupied was the same as one atom will claim in our present day and age. Normal gravity started at 10^{-43} seconds. The Universe was the size of a neutron or somewhere in that vicinity. The Big Bang began and GUT, or the grand unified theory, produced the attempt to describe the strong and weak nuclear forces and electromagnetism in one single mathematical theory. Somewhere before 10^{-12} seconds of counting the Universe cooled to about 10^{15} K the electromagnetic and the weak interactions acted as one single physical force. Science reckon that unification may come about at temperatures of 10^{27} K, which was the temperature of the day at 10^{-38} seconds after the Big Bang. This statement echoes my viewpoint but one has to look carefully for that to surface.

In the suggestion the presumption claims that all the space that the Universe made available at that time was the total space one atom might take up today. If that might be the case then where was the rest of the space that now fills the Universe? Or was the rest of the space we now find in the Universe and what is now explained away as the vacuum, also available back then. Did the Universe only have that one tiny hot spot it filled with huge volumes of heat? Was the rest of the space vacant being out there all along during all the time running to the present date but filled with emptiness standing around as a big vacuum with no better to do than sucking on the Universe while the Universe was exploding at the speed of light.

Then that statement suggests that in this hot Universe there were light-years upon light-years of vacuum waiting to be filled by the intense heat soaring in the smallest spot. If that is the case then why did the vacuum not fill in the blink of an eye by all the exploding expanding material growing at the speed of light? Was the Universe overall bitterly cold where the vacant space was locked in with one spot of the vacuum filled with temperatures so hot we can only produce it in numbers suggesting a value but never claim to be able to digest the reality thereof in the human mind? If so what happened to the natural consequence that heat flows in the direction of cold and equalise between hot and cold. Was the space being available at present available then or was the hot space the only space available at the time. If so what prevented the heat from instantaneously filling the eternally cold vacuum because with the rules controlling vacuum in affect, it should have filled in such a manner in less than a heartbeat?

I believe that singularity formed space-time and space-time developed from the overflowing of space-time at the time is extending by marching onwards and outwards to this day. Space-time developed another product that everything in the cosmos has to have. It must be in such large quantities everything imaginable in the Universe has to have it and that is space using time to move about. I suggest that it is space that is holding heat in a quantity providing density and ratio to space available and in relevance to the space being available to quantify the presence of the heat and which then proves to form the time factor. The container and contained all together mixed by motion. From that very first separating of heat and space, which is what formed from singularity to produce space-time. The Universe was full… It was overflowing by the speed of light in the beginning…so where and when did vacuum or nothing enter the Universe as a factor if and when the Universe was so full.

The answer to that is absolutely crucial because how did the Universe decide to fill some parts with a variety of something and decide to fill some parts within the in-between with nothing? If that is true why did gravity not prevent the vacuum filling because no gravity that came about since can beat the force that gravity had back then? This leads to another question following the previous one in asking why did gravity at the time when it was so strong with r^2 so much compromised not fill the nothing immediately as it entered with something that could absorb the nothing. At the very beginning the mass that was pulling on the mass by force was immeasurable and none quantifiable. Even more to the point is the question to be asked in how big was the radius between the materials with the immeasurable mass placed in such a little space. This is all the more important in the light that the smaller the radius is the bigger the force will become from the immeasurable mass pulling...With the immeasurable mass that was producing the first gravity between the particles divided by an almost non-existing radius the gravity produced had to be in gigantic proportional quantities and with the separation of the radii being in the infinite measure that it was at that point then how did the Universe establish the chance to expand. It did expand, as we all are witnesses too in spite of this contraction of gravity that had to have been compromising the expanding factors.

Still the expanding filled the unknown part of the unoccupied Universe, which at the time was there or was not there and it was there it was then filled with nothing. If the nothing was not "nothing" then the nothing that was not being nothing was also filling the rest of the vacant Universe that was or that was not because if it was it was filled with nothing and if it was not then it was nothing. This is then taking into account that then all the reducing that is resulting from Newtonian contraction and that was going about in the space available at that time was something filled with nothing and surrounded by more nothing? With everything in the Universe being that much crowded and crammed where and how did nothing enter the Universe and fill the rest that was unfilled? What factors introduced nothing into the picture since the entire Newtonian concept finds its base on the principle that matter reduces using gravity by force which then bring about reducing or the removing of the many nothing between particles, which will then lead to nothing that has to vanish even before nothing can enter the space. This question may seem small-minded belonging to the mentality of a child or to that of the mentally impaired with not much factual appreciation developed yet. Please do not see it that way. If you think in those lines it will be because you do not have an answer to challenge these silly questions. Beware, silly as they are they represent official backing by the Wise-and-Informed. If the space is nothing and if the space was as large as it is at present then there was no need for such a small area to fill with something leaving only the rest filled with nothing at first since all the space we know about was there present and by being present it was there then for the taking. What ever filled the Universe had to start at the centre of the Universe and fill the entire Universe all over from a centre as it moved outwards filling from the inside outwards.

This is a natural human instinct realisation but is beyond proving by using Accepted Scientific policy. But that leaves Newtonian science with a massive unsolved problem: where is such a centre at the present time and where does the centre produce the limits or border it apparently has to form as it expands. By expanding there is an additional contribution too that that was when that that was, was receiving more than there was before the addition increased that which was and by then becoming more than there previously was it had to be improving the border from where it must have been before the adding took place to where it was after that that was added was added. When that was less than it became when it was added too it was at the limit that was there before it was added too and that limit there was, was a limit that is the limit that I am referring to as a border being there. The cosmos is filled with unrecognised borders. The expanding has to be an ongoing filling that is at the same time expanding from the inside towards the outer limits of the Universe. Since nothing can enter from the outside where nothing is, the filling of nothing as a substance that would take up vast quantities of room had to fill from the very centre spot where all other filling came from. This filling of nothing with material has to be well mixed.

The truth about cosmology is that space forms no borders but by using any Newtonian centre from where mass is attracting we must find a point where there has to be the ultimate Universal centre which is the cardinal point in the entire Universe and it is the first, the prime position to locate coming before any other concept one wish to put forward because all concepts has to start with locating that cardinal centre. There has to be the ultimate r^2 radii located precisely between the ultimate mass drawing the other ultimate mass closer. If there was a Big Bang then there has to be the spot where from the Big Bang developed therefore there has to be such a centre connecting the past to that ultimate centre with the line of development flowing onwards to this day. The fact that science is Newtonian proves that in the meantime Mainstream science is still of the opinion that there was the specific centre in the Universe that is nowhere to be found as it was filling the unknown with nothing coming from nowhere, but which somehow is still somewhere in the centre of all of that which is something. On the opposite side of nowhere there is an outer border in space producing a limit to nothing and serves nothing with a specific point to stop being nothing because that point is precisely where nothing ends and forms a beginning of a Universal border or a Universal end. How one will stop vacuum being no longer nothing was a question everyone comfortably missed to ask therefore no one ever seemed to deliver any form of answer.

One night some years ago very close friend of mine had a meal at his restaurant and as the conversation progresses he asked me about space and where it must end. I tried to explain to home what I believed in comparing to what Mainstream physics believed but soon saw I was not gaining in his understanding. Then I decided to jot it down on paper and he could read it at his leisure as he saw fit. That led to the first book written by me (in Afrikaans my native language). What I tried to explain to Johan Boonzaier that night is that if the universe was the size of say even a tennis ball with only the size of a tennis ball being the very all of space there is available, then yes, it must take time too expand from that having the excessive heat there was back then in all the space we have at present. It then is converting heat into space bringing about the expansion. But one will find most expanding within the atoms, as the atom must grow since the Universe in all was the size of what one atom is today.

The space in the atom pushed the space outside the atom but there must be plenty more too the growth. Something outside the atom contributed in it own rite because there is more expanding than there can be blamed on coming from the atom. But the space then also developed as the universe developed and if space developed then it cannot be total vacuum filled with nothing because "nothing" cannot develop. You the reader must judge whom is correct between my view that space developed with the Universe as part of the Universe and reject the official view about space being nothing or otherwise you the reader must then decide that I am wrong, but should you do that, then find a reason why the Big Bang started out small and filled all the available vacuum or what is contemplated as vacuum that we have with the motion of time. When Mainstream science accepted the Big Bang as the principle that will take science into the future the view about such a Big Bang concept unlocks a different door to another view on the cosmos from birth to end. It calls for revising all aspects of the entire history on cosmology and change what dead wood needs chucking out. Most of all it was my following the lead I got from Kepler that unlocked the doors I now present to you. I claim there is no graviton as there is no gravity forming weight or forming mass. I hope the sketch helps with my explaining effort:

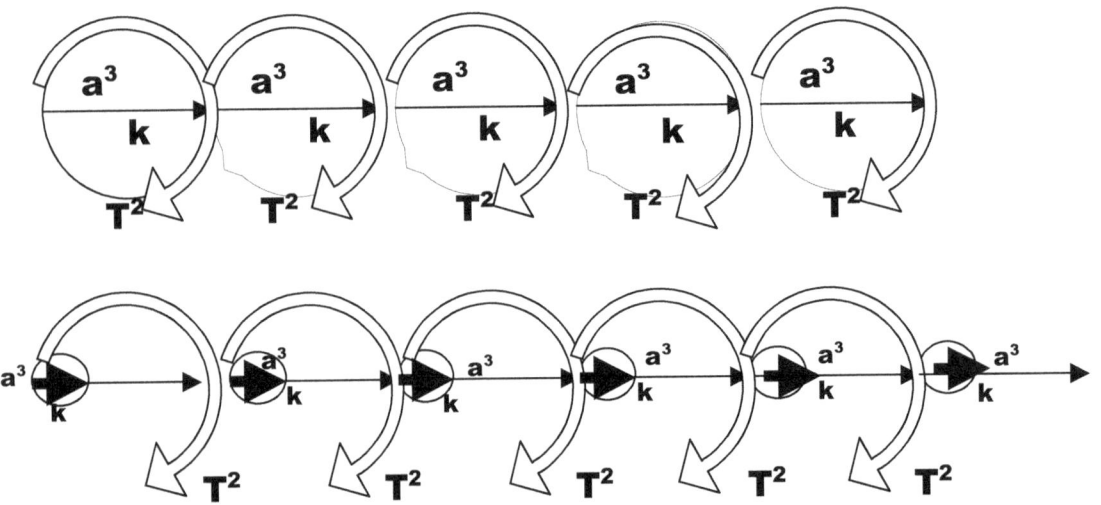

Two objects of substantial size differences are travelling at the same time but one has a space, which it has to move when it travels that is considerably different from the larger space. The larger space will produce an extending line equal to the space it moves while the smaller space will also produce a line in ratio to fit the space holds relevant and that it has to move.

⟶ ▶ Is the line that $\mathbf{a^3\,k}$ representing the larger space has to use to duplicate while using $\mathbf{T^2}$

▶ Is the line that $\mathbf{a^3}$ $\mathbf{k}$ representing the larger space has to use to duplicate while using the same time constraint $\mathbf{T^2}$

⟶ ▶ The difference there is in length brought about by moving the larger space in the same time $\mathbf{T^2}$ as the smaller space is what brings about mass. There are other factors too which I shall touch on as the book develop.

Mass has precious little to do with the whole affair except to be an obstacle intended to restrain the motion of the hosting space. The difference in size between the one in circular motion and the space in contracting motion must bring about that the smaller object has to move about a circle much closer to the centre because the larger space form the centre hosts it. However there is no large or small in the cosmos but only those better developed or those poorer developed.

By duplicating there is more to duplicate in the better developed than in the lesser developed. When the lesser-developed space is duplicating the less developed space would hold a lesser extending from point to point forming a shortfall by distance in comparison. The motion being extended needs less extending and should therefore be closer to the centre in relation to what the better developed space would need in extending by a duplicating effort. This is the principle we find that is behind the sound barrier. The motion the aircraft produce forms an increase in the duplication of the aircraft which extends the duplication splitting the Earth producing a extension and the aircraft producing an extending that goes beyond the attempt of the Earth's extended by such motion.

I know this may sound barely believable but please hear me out. While we use gravity the use of gravity as such makes us part of the Earth. We see gravity as some influence or force producing mass and that mass is forcing us down on the solid ness and onto the Earth. By having the mass we become a semi unit with the Earth. That is how we on Earth see gravity but when investigating gravity in outer space we must come to a basic question: Is that what we experience as gravity on Earth truly gravity?

Much of the proof about gravity is part of our perception about gravity because we experience certain conditions with gravity while we find ourselves bogged down on Mother Earth. But are our perceptions about gravity truly correct? We experience mass but are the mass the result of gravity or are the mass the product of gravity. We only experience gravity, as a factor from the position we have on Earth and the conclusions we form is a product of a perception we formed while we are being forced to be part of Earth. It's as if we are upside down and has to decide on which route we should follow. I want to make a suggestion, which I aim to prove in the following pages. My personal being on the ground and having mass that is keeping me on the ground comes about because of the speed that I travel through space being the very same as that which the Earth has.

By me not applying a speed difference I then inherit the speed the Earth places on me. But my space which I use $\mathbf{a^3 = T^2k}$ to travel and the space which I use tot ravel through is much smaller than that which the Earth burdened with to move and to move through. By me having a smaller space to move $\mathbf{a^3 = T^2k}$ the space $\mathbf{a^3}$ being moved $\mathbf{k}$ in the time it would take to move $\mathbf{T^2}$ will produce less space $\mathbf{a^3}$ to shift $\mathbf{k}$ and therefore a smaller distance $\mathbf{k}$ to replace all the space a^3 that is moved in the time T^2 the space a^3 needs to enable it to move **k**. To duplicate by motion the smaller space requires a smaller distance to shift the space but the motion will take up, as much time to complete than would the larger space take to complete though the space the larger space has to duplicate will require a longer distance to complete the total duplication of the larger space. A large space a^3 will produce a large extending **k** when using a^3 the same time duration T^2 when using the same time factor as that which the smaller space is required to use when under obligation to use same time constrain. Behind this is the most basic principle hiding which allow us the fortune to be able to fly using a flying machine. It is all about motion supplying relevance and forcing on time constraints.

Because my body that I have is travelling so much slower than the Earth is travelling due to my size in relation to the size the Earth has and although I am using the same time as the Earth does to move, such a speed difference is not in the time differences it takes to complete but in the space differences that has to be completed in the same time but is unable to fill and the space is trying to crush me into the Earth where I am forced toward the centre. If I were able to penetrate the soil solidness I would reach a point where my speed as zero would equal my space I occupy.

The space I duplicate by moving from one position and placing the space I hold in the next position while keeping my space I move as it is identical in the next spot but located in the next position. Such moving by duplicating takes a certain time to move from one spot to the following spot and it will use a certain frequency that will have the same ratio in bridging the gap from one point to the next point as that which the Earth has.

My speed of duplicating by motion has to be even in frequency because I am within the duplicating space, which the earth is duplicating and as part of the space that the Earth is duplicating but the duplicating of my space I do myself. But in size there is a massive difference between the space I hold and the space the Earth holds but to duplicate will take me as long as it takes the Earth. Notwithstanding this common factor the Earth has to use equal time in duplicating its massive space, as I have to duplicate my small space when we both have to share a frequency that will keep us duplicating evenly. Therefore the frequency of duplicating using the same time period will be a lot different to my much shorter frequency of duplicating space.

The difference is between me being in mass and me being in the correct position in the space-line the Earth has will place me in the correct position but the heat that then will surround me will fry me into non-existing. Fortunately for life the soil forms a barrier through which I cannot fall any further as to correct my location. But being where my position would have no mass would allow me to float there in that location in the same manner as I would float in water. I would be buoyant. It is because I do not harmonise the displacing frequency as I should that I have mass.

My having weight is what Mainstream Physics use to give me my gravity. Science purposely switch my having mass and confusing my mass with my having weight to explain what is beyond explaining. It is said that while I float in outer space in state of suspending hanging above the Earth in the weightlessness I still have all the mass that I had on Earth. Butt in order to prove that those in science will give me a mass even in outer space whether I deserve it or not. By that token science first have to cheat all logic by reasoning in some bazaar way that I take my mass up there to where there is only micro gravity. They firstly claim that all of a sudden I take my mass to outer space and in their next argument they say I have micro gravity in outer space since my body is floating as if it is in the sea. But if I stop floating and start falling to the Earth my body and I did not gain any mass. My falling then comes as the result of my motion being much smaller in relation to the space I claim and my motion then is being less than what is required to keep me in the position I have which in I maintain my orbit up there. By moving to slow I fall. I do not fall because my mass grew. But science has been proven wrong by their work without any of them aver admitting to such a defeat. All the satellites fall if the satellite motions are not reset. The satellites do not gain or lose mass. They gain or lose motion

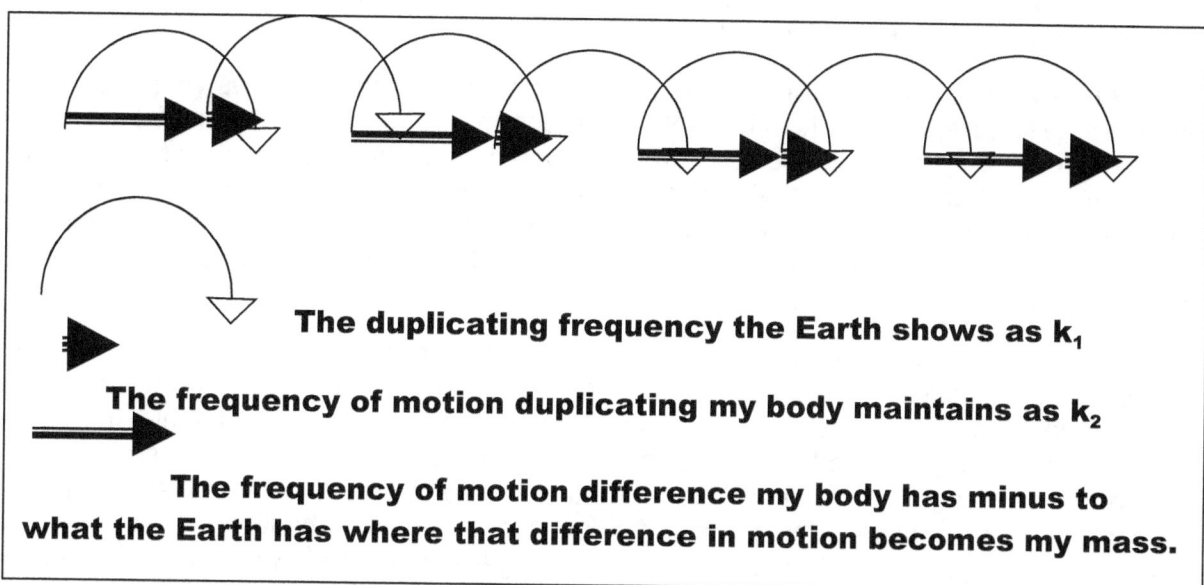

The duplicating frequency the Earth shows as k_1

The frequency of motion duplicating my body maintains as k_2

The frequency of motion difference my body has minus to what the Earth has where that difference in motion becomes my mass.

By amplifying either my space (using a Hot Air balloon) or by accelerating my motion that I have in relation to that which the Earth forces me to have, I will break free from my weight or mass. I shall become airborne and float as if I am in outer space. By pretending my mass can be multiplied many times over in using a process, which then is called not gravity but momentum. But motion and gravity is all the same because motion is gravity that is redirected, which then forms another part of gravity where gravity again is also only motion applying. Science maintain the argument that when I am in outer space and am no longer part of the Earth I then will only have mass. But since there is only micro gravity I will be in a state of weightlessness.

My mass is what gives me gravity and while being up there I take my mass long with me. But with my mass up there I will only have micro gravity. I am floating with my mass and it is my mass that is responsible for my gravity and I am floating above the mass of the Earth, which is rite down below me, but still I have micro gravity. That is true if I wish to incorporate the dubious use of double standards by separating mass from weight. The mass my body will have in a Black hole will be a billion times (at least) more than what it is on Earth. With that the Black hole destroys the fact propagated in science that my mass will be the same everywhere. That is more than permitting double standard. Because our motion is much slower than the earth is spinning we place a breaking effort on the velocity the Earth has and that breaking effort we accept as the mass we have. The truth is that my mass comes about from the lack of motion I have in relation to the space I occupy and has nothing to do with any gravitons pulling me down. If I increase the motion I have there shall come a point where my motion will be sufficient to pull me into the air, as I then will have the required velocity to lift from the ground. That motion being in excess of what I have and is complimenting the motion that I receive from the Earth counteracts the motion of gravity that is containing me.

The motion I adopt then release me from the motion containing me and if motion can release me by only becoming more then gravity is my motion not being enough in the first place to keep me onto the Earth. Nowhere and at no time does my mass ever gain by having more protons that will get me back to the ground as if I am bigger or carrying more material or does my mass reduce to get me into the air as if I am smaller or carrying less material. Please note that this is my way of explaining to you about the fact of bodies having weight or mass. It is not mass or the lack thereof or any means to measure occupied space within the atmosphere of a larger body that pins me onto the ground. My body is claiming space by motion in space. Gravity is the result of motion because it is in the motion that bodies have that gravity affects them. This is proved because by adding motion the mass does get more but the body never gets bigger or hold more material, and in defiance of that statement by increasing the motion my body lifts and flies. The reality is that my body in motion has more mass being momentum but still my body lifts when motion allows my body to lift. This statement confirms Kepler that a^3 becomes more (massive) when motion T^2k becomes more (moving).

Mainstream physics admits all along that nobody, human or otherwise knows what gravity is. While investigating Kepler's work with employing much motivation and detail in order to give his work the much duly credit it deserves it will also serve a valiant purpose when by the same token we try to establish what gravity is, because I believe Kepler possibly answered that mystery. We have to start with the person that introduced gravity or so does everybody acknowledge. Newton saw an apple fall from a tree and he subsequently realised there is some force pulling the apple to the Earth.

Allthough he still was a student he announced his findings and became a genius on the spot. The concept he introduced as gravity gave him instant admiration from which he became the legend he is today and that reputation he gained there at that moment would last him from that day he instantaneously unveiled his mastermind, and that same genius still serves him in his honour to this day long after his death. He found that this force has to have some thing to do with the weight and the mass of that particular object and the mass of the Earth. There is some force pulling that apple as much as the force is pushing the apple and the same goes for the Earth because the mass the Earth has is doing the same to the apple.

Between the two objects facing gravity there is a force that develops where such a force is pulling the apple on a constant basis towards the Earth even after the apple is already in a steady state on the Earth. That forms the mass and the mass forms gravity. He concluded that the mass is responsible for the pulling. Remember this observation came three point five centuries ago when knowledge and brilliance carried a much different defining than what such defining of brilliance is worth today. He realised the pulling on that apple brings about weight that brings about mass because the apple departs from its location and arrives at its end location when the falling is completed. Then he went out convinced all that was in line of finding the needed convincing because no body before Newton thought of what Newton thought quite in the way that Newton thought about gravity.

Newton succeeded because he found a way in presenting science with the fact that objects move closer because of some force. He went one step further and named the force he fathered as gravity. But there it stopped! Any and all other further defining the matter or going into any possible observations of whatever magnitude concerning the topic never realised any motive to go further. Inspiration to further commitment just flew out the window as the essence to do so immediately expired as far as the rest of science is concerned. What might he have missed if he missed anything? We all fall down when we are unstable and out of balance. He never realised that balance is more crucial than brutal gravity because that part is the defining part about gravity. No one ever gave a thought about the balance part even centuries later even as we grew into all the sophistication we now enjoy. What brought about the balance that secured objects in an upright stance and supplied some form of control over the managing of a position?

Any other position than being flat on the floor would have a better defining than being just at the mercy of the force gravity. Standing tall is a stance that defies gravity so there is another force other than the pulling of gravity. Admit tingly the force would first and foremost have to aspire to the rules of gravity and then comply with other demands. True enough is the fact that that position would ultimately and firstly by all accounts have to satisfy gravity before any further motion could commence. Yes but then by balance motion defies gravity by changing gravity's force of pulling everything straight down towards a visionary centre between the objects. In affect this means somehow there is control over gravity and gravity does not leave objects beyond outside control. Gravity is manageable and can be controlled; we just have to find a way…

Years later some one came up with the novelty of hot air ballooning. Ballooning proved that there is antigravity but that part was missed by all even to this day. Some people speak of antigravity as if that is some mystifying mysterious concept that is so well hidden in the secret annals of the hidden Universe that only Ali Baba and his magic words can reach it. Please consider the following statement. If gravity was bringing the object down, because of the affect of gravity which is that what we experience as the gravitational sensation and that is what we interpreted as gravity by our sensation and observation, then that is only coming about by our bodies that is in a state of being dragged down. The dragging down of the body is in the direction of the Earth centre. That sensation of being firmly locked onto the ground constitutes to what we believe we experience as gravity.

When some influence bring about the very opposite affect, which then results in establishing the opposite result it deserves to be anti. In example we feel dragged down but anti will be the lifting of the body into the air. Anti will be going in an opposing direction of the motion that gravity inflicts. It will counter the influence that gravity apples. Such motion has to indicate antigravity. The counter acting of the mass dragging us down must be anti gravity pulling us up into the air above the ground. Antigravity must come from such an opposing influence that will bring about the lifting of my body. If hot air ballooning gave the object an opportunity lift, then ballooning must be antigravity. The balloonist and the entire balloon found a manner to counteract the pulling of gravity enforcing weight. The balloon can lift what gravity depress and if Newton said gravity is the falling then later Newtonians must agree that the opposite of falling is flying or lifting. A balloon is lifting-and- flying. If gravity is pulling down objects in the direction of the centre of the Earth then flying is antigravity. Moving away from the Earth by means of motion and in particular flying is using whatever means to defy gravity where the lifting can also be the hoisting of a body by a crane. Lifting by ballooning in a hot air such balloons escape from gravity where the balloon constitutes to bring about the effect of establishing antigravity. Climbing up mountains must fall into the antigravity department because parachuting down the mountain definitely falls in the gravity department. Nevertheless it still does not answer the question of what gravity is.

Let us look at antigravity because the antigravity is releasing the object from the gravity that controls the object by an Earth fed force. The balloon starts flying when the confined space of the balloon is veraciously and violently heated in access. The balloonist shows us that in order to overcome gravity we have to introduce heat. That is the only manner in which we can defeat gravity. Even by an engine driving an aeroplane such flying can only result if an engine combust solid fuel by creating motion as the fuel mixture is turned into heat. It is heat that makes the difference. That is the very thing that Kepler said. Expand the space a^3 and the motion T^2 will move further increasing k. Blowing hot air into the balloon is increasing space within the balloon a^3 which then results in providing the balloon with a larger distance k from the Earth centre k^0 that still holds time with in the Earth atmosphere with the Earth T^2 within the space of the Earth k. Using Kepler provides us with insight and the ability to see what gravity is by showing us what antigravity is (a^3 gets bigger and that will bring in a larger k). But moreover the larger space in enough compensation to bring about extra motion that will defeat gravity by the extending of k. If that is not antigravity then we can forget about Ali Baba and his magic rhymes too.

The balloon assists us to escape the Earth's hold on our body, because there has to be the force producing motion countering the motion of the Earth gravity. The balloon shows that releasing enormous quantities of heat into an inclusive area excluding space such as that which the balloon canvas provides, which is establishing the release from the gravitated containing force on the body giving the body a means to escape by floating about above the ground. The motion is at that point breaking free from the containing gravity by moving in a specific direction, other than the direction the Earth gravity inclines the body to travel. By concentrating the releasing of heat into the balloon, the direction of motion starts to contradict the enlisting of the Earth gravity and the heat breaks the balloons confining properties while the balloon is released from the Earth as the balloon and us lift up into the air and away from our confining to the Earth.

At the point of explaining we arrive at the point where we can say what we think the difference is between the balloon floating in the air above the Earth and a body suspended in outer space floating above the Earth's atmosphere. The difference is the heat that is in the confined air per volumetric ratio favouring the heat being more in the space than what the heat is outside the confined space. If we had any method to put the required heat we need to escape from the limits of the Earth to outer space into the canvass of the balloon there was no canvass left to contain the heat. The heat is available to do the job but the means to do the job with the tools in hand is unavailable as far awe can use the balloon. By having more heat in the one area than there is in the other area beats of the pulling of gravity. Obviously it is antigravity that keeps the balloon in the air and what keeps the balloon in the air is having a larger volume of heat per space unit than what is in the atmosphere. The balloonist shows us that by applying more heat we can defeat gravity more. Someone took the advice, because the next minute the Germans had rockets. The launching of rockets brought about the ultimate defeat of gravity but it involves almost the ultimate releasing of heat.

In antigravity we find heat more concentrated in one definitive area than the heat concentration is elsewhere. The more the heat is that we release into space the more the antigravity is that we achieve and the more release such antigravity can produce. But what connection can gravity have with heat and if there were any connection between heat concentrated and gravity, what would such connection be? The history behind Carl Benz should bring the answer but more so would be the story behind James Watt and steam although the James Watt story may not be that thought provoking because it is much less filled with the ever popular cheap thrill only sensational gossip can provide...Still both stories cover the same principles. In the Carl Benz story a housewife leaves a pot of benzene fuel on a coal stove. The pot with benzene heats up where the pot with benzene becomes hot and under pressure. This performing of heat increase releases the heat as newly creates space, which then removes the housewife with her house from the neighbourhood she used to regularly frequent as her residential address. Afterwards almost the entire neighbourhood is not there to tell the tale or ask why...

It was a stupid tragedy that brought about the end of steam and the rise of the internal combustion engine and on Earth billions on billions of human souls are in torment not to please or suffer for the advantage of coal Barons any longer but now they are dying and suffering in agony to please the wishes and desires of oil Barons. How much did the world not change...While it is no longer the coal Barons shackling us in chains and telling us democracy broke our burden of slavery, we have now the pleasure of the oil Barons enslaving us with democracy and telling to be happy because we are the fortunate slaves, there are others circumstances in which they can enslave us that will leave us worst off. All this came just because the pot of fuel created a houseful of space that was enough to remove the house from the address the house previously enjoyed. But Mainstream science neglects to appreciate this. They see the heat, they see the antigravity but they fail to add the heat, the anti gravity and the space that no longer housed the house of the naive and rather impractical thoughtless housewife. They call the tragedy an explosion but then again everything that expands while using a noise during the expanding is an explosion. Adding of new space to the space holding the house at first altered everything that was previously proportional positioned in the space where the house was. Such exchanging of heat to accumulate and introduce more space in the process referred too as an explosion was bringing in more space that came directly as a consequence from the explosion which was producing more space where the increase in space brought disorder because the well organised material distribution and placing was before the event filling just enough of the required space arrangement that was holding every object in a prearranged order of tidiness.

Then suddenly out of the blue the space which held the house in a tidy arrangement had to accommodate more space therefore the ratio of material per space volume increased dramatically many times over in the favour of the space in the balance. That part no one ever acknowledges. However the losing of the house was not much surprising to Mainstream science back then and even today because who cares about old news. All of Mainstream science was at the time as they are today very familiar with all explosions because of wars and bombing that leads to maiming and killing and all the unspeakable monstrosities we associate with war so that the dirt poor can suffer and die to leave the disgustingly rich even richer. The poor has not the means to pay science to be clever and devise methods to save their lives so the rich does the poor the favour of paying science to find methods whereby more poor could be killed as long as the rich saw it as a good investment with great capital gain on the part of the rich. Therefore science is well established in the method of creating more elaborate and destructive explosions that the rich pay them to invent. In the explosion caused by our housewife no one put up money to investigate what happed during the explosion but money went to why the explosion happened.

That inspired an investigation in connection with the fact of the finding more about what takes place during the carnage as more money goes to finding means to create more carnage per money unit spent. At least that is why the poor were invented and that is why wars are invented. It is invented so that no money goes wasted on saving the poor people except if the poor has the money ready and available to pay the rich for medicine to enable the poor to stay alive. So science goes out and develop more fuel for carnage but fails to find out why the housewife and her house is no longer part of the neighbourhood she use to frequent. With the loss of the presence of the ignorant housewife with her house her neighbourhood and all was a normal way of leaving us with a new way of tapping and harvesting energy and untold riches which was born with the death of the absent minded housewife. But according to the mindset of science they saw not what the incident presented in space producing for to their view nothing new came about since it was just another exploding of fuel...so no body bothered as to finding out how. What they missed was the part that the coal stove played in the whole tragedy. Without the intervention of the coal stove producing the heat that turned the liquid fuel to liquid heat liquefying the space that turned the liquid space into a gaseous space where the liquid space revealed its true incentive in nature by turning out as space and the newly created space that was in fact liquid space that went onto become more space, well that space was providing the one main factor in space-time relevancy.

The stove's heat producing space by transferring heat leading to the expanding of the fuel as such expanding was creating new space that is transforming all other surrounding space and is rearranging every aspect that contains space or that space contains. It will bring a much different looking end. Everything about this concept is missing from Newtonian science because Newtonian science failed to investigate Kepler. Kepler said space a^3 is equal to the motion T^2k thereof and then that says without Kepler directly saying it, it says that if space a^3 goes bigger as a result of the explosion then such increasing in space will constitute to more space a^3 which has to produce an increase in motion T^2k where more motion T^2k will bring about faster displacing space. This is one small fact that Newton robbed the world of realising with his ignoring of Kepler's work.

We are now serving time in the twenty first century. One Professor once told me I must realise that Newtonian science took man to the moon and back several times and in such a view I am rather annoying presumptuous to criticize Newton. The Professor missed the point. I criticize Newton on what he did not give us, which he gave us as incorrect by his own admitting that it is mostly guesswork on his (Newton's) part and his guessing about the facts where later that guesswork became institutionalised facts believed by all concerned to be correct and to be proven to a degree of correctness that is far beyond doubt.

Newton gave us gravity but Newton never gave us the explanation about gravity. At the time Newton met strict opposition from his colleges and piers because others felt his introducing of an unexplained force was taking Science back in time, which of course it did. Many scientists at the time accused Newton by name of dragging science back in the wrong direction of progress by introducing unexplained forces acting in a superstitious and mediaeval manner.

I went one step further by asking myself the question: If space becomes more when heat becomes uncontrolled why can space not become heat when space is under control? If space becomes more as we see with every explosion of every kind and such heat forming space releases energy, then why would space being managed not form heat being under control and produce energy. We only have to see what Kepler said gravity is. Motion gives us energy.

Where space is the least, which is in the centre of the circle gravity is the strongest. The gravity located in the circles space less centre holds not only the sphere together but all that is in the surrounding of the sphere outside the sphere as well. It is from there in a giro action that gravity bonds all atoms forming the structure of the sphere as one unit together in a unit as well as distributes a specific alliance in shape and form. How the atoms manage that we will get to in a while, but there is a law allowing for that to take place. Gravity is the strongest in all cosmic structures holding the form of the sphere and gravity controls all around from that very centre where space is the least therefore the more any star produces gravity. The smaller the star is as far as volumetric occupation goes, the stronger the gravity is that is coming from such a centre. The less the space there is the less the motion is and therefore the stronger and more deliberate the motion is evoking gravity. From the centre in the middle where space is absolutely at a premium the gravity grows stronger as it draws all material.

The motion is one of confining the space to a centre by the moving or trying to move the flow of space and whatever is in the space into the centre where the space is least. Take the Neutron star and the Black Hole as an example and compare that with the sun and the answer is self-proving. I claim that gravity is all about reducing space and not attracting matter but that I explain a little later on. Therefore the matrix of gravity must be permanently located in the location where space is the least. Looking at a sphere we find that what holds the sphere true to form is placed in the centre of the sphere, which then has to be the most intense point of gravity. Gravity is confirming the round shape without favouring any specific point. Such evenness of gravity come from what is applying at such a centre and is in control of the surroundings. The centre that secures all of the space and material in the space holding the specific form has to be round if it is anything. That shows that in the sphere one can see that the sphere as a form is dominated or controlled from one specific location in the centre. The explanation about the reason there is control coming from the centre has a very childlike simple answer.

From the every every part of the matching and side of the circle circle. Between the line connecting the two **All connecting centre of individual connecting lines between opposing points** point there may form on the outer circle line of circle structure and all structural positions of the circle in all circles, all circles refer to the centre in perfect aligning. As every point wherever located on the sphere has a equal but an opposing point on the other but in equal position on the other side of the two controlling points runs a precise straight opposing points in counter balancing. When drawing the connecting line between the two controlling points and connecting such points on furthers edges of the circle by lines formed the lines will all cross the centre. From wherever a line may cross and from every point forming a line to the other side of the circle rim holding the connecting points there has to be a counter point located on the very opposing side that when connected by a line such a line crosses in the centre. In the middle the centre spot bonds all sides coming from any and every direction there can possibly be. The line will run to an equal point on the other side across the same distance from such a centre and that then has to be where the strongest gravity can be located.

The Big Bang was where gravity held the Universe in the least space there ever was. To find the original gravity we therefore have to reduce the sphere to the circle and reduce the circle from there narrowing the circle down to as far as one can go. The Universe is a magnitude of spheres constructed by a complexity of circles. This is because everything sprouted from on matrix singularity. To narrow any circle down will be the same as narrowing down the Universe. In our reducing of the Universe we must first acknowledge that the Universe constitutes many spheres, which is giving the Universe gravity as a combining unifying part which is the part of the sphere giving the sphere form (or gravity) and that confirms that the sphere is a circle in many times over multiplying the positions from where gravity secures form. If we wish to go back in time by taking the Universe back down the same route and at the same time maintain some coherency we must concentrate on a single circle because a sphere is a circle by millions of possibilities linked together by just a name that changes the concept.

When one takes this accepted route in thinking that by reducing the connecting line to the connecting circle point in the centre of the lot, it must take us back in time at the same time as the circle reduces to the time during the Big Bang. During the Big Bang where all circles was as small as they can get we run into an unknown substance we came to know as antimatter. This theory is propagated according to Mainstream science but what is most surprisingly I do agree with this part of the statement. All material produces gravity. I go one step further and say all material apply motion where some motion may be to contain by using gravity attributing to the contracting that leads to the reducing of their space.

Then as everything in Creation has an opposing the restore and maintain balance, there had to form another or other material that did not by our lamentable standards produce gravity because those material produce antigravity, a concept beyond human discernment. Antigravity must be the expanding in counteracting contracting. A counter action to contracting is where expanding provides pappy to that which has no gravity. Forming pappy provides more space by losing density to the advancing of their space. Material either have gravity by solidifying or concentrating the space they hold in ratio to the material within the space they hold whereas others lose their solidness by entertaining more space within the ratio of material to space where such material becomes liquid and in more extreme cases they become gas. Being a gas they float which gives that material a high degree of antigravity being airborne. It is however not clear if antimatter produced gravity as it did when it went to lunch on and ate up all material in the immediate surrounding. It was cannibalistic but the unanswered question is this: was it a gravity producing predator or a non gravity-producing carnivore. Did material find a comrade in their gravity forming of form or did the gravity it produced bring on the demise that subsequently followed the event as is reported by the highly informed.

The Accepted statement on antimatter reads that matter composed of anti particles where such subatomic particles that have identical rest mass to corresponding particles of ordinary matter but opposing charge and are opposing in other fundamental properties. One example given is that an electron would have a positron, which then functions as the anti particle and has a positive charge compared to the electrons negative charge. That is put bluntly in its utmost simplistic form. Unanswered and tough questions arise from such a statement. What kept the electron bonded to the atom since the protons must by implication produce expanding or by definition be repelling the atom and surroundings instead of the normal contracting or confirming of form.

What is a positive compared to a negative charge, because it is human concepts that put the directional qualities of material into a positive or a negative contexts as we did with hot and cold. It is human standards that humans brought about to make all human inadequacy by lamented human understanding better but it is not applied cosmos principle. If there is extracting electrons performing in the capacity as antimatter, then there better be protons by other name in service to the anti electrons, which then of course serves the anti electron in the capacity of an anti proton with an equal but negative charge to that of the proton. When matter and anti matter meet the two opposing particles annihilate each other until one vanishes from the universe. I have to add that at the time this theory was devised the first computer games became a crazy fashion played by young and old, those wise and those foolish all alike. This game was called the packman and the packman ate up all the skulls and after eating left nothing as evidence.

The theory about antimatter has some very striking similarities to that packman game. It still does not answer the most ardent questions: What makes a positive electron different from a electron in the working place each has and can any person show such an object found in nature. Can people take a positive electron to an investigative bureau and are awarded for such evidence? It is unwise to substitute nature with human concepts just to further mathematical equations. This was apparently presented as normal as nature was when nature developed with the Big Bang and nature then did behave this oddly just after the Big Bang came about. But one huge misgiving in this argument is declaring that everything the antimatter had as a meal vanished and even moreover then antimatter went and vanished too.

Where could the combination that was produced when the matter and antimatter collided go after it disappeared and did it form the by-product of antimatter science is talking about, which since then apparently vanished too.

What a bloody none-intellectual fairytale that is on the bargain too one of those made–up-as-they-go-along stories, which is told by persons that supposedly should know of better. Since there is no place other to find a location to be within than being in a place inside the Universe it is hardly possible to vanish from the Universe except in fairy tales because for one simple fact: there is no other home to have but the home we have which we call by the name the Universe and we have no where to escape too but within the walls that the Universe provide for such a purpose.

There is one Universe containing all and preserving the lot. Mainstream physics is accepting this fact. But then by the same margin they accept a principle that allows property that once was part of the Universe to leave the Universe and go somewhere outside the only Universe. They create a loophole whenever it suits them to misplace what they cannot explain readily and logically. In Creation to their and my thinking there can be no hiding of anything but in the Created Universe. This they admit and confirm although with the same breath those very same intellectuals also admit that there is another place outside of what we are able to find in the Universe. When someone comes up with the marvel where such a person can declare in all honesty that the product of antimatter or singularity escaped from the Universe to God knows where that person should leave the field of science and go for fantasy writing such as fairy tales or reporting about politicians inner deepest chastity and integrity. That is what we can find outside the spectrum of what the Universe can deliver.

With such a statement of any Universal product disappearing from the Universe alarm bells should go off in the mind of the trained and professional Scientist working with such matters. Yet those in charge do not once belabour a question of the validity of a statement that involves a stating of factors declaring the possibility that there was are now an outside of what once was part of the only place there ever can be. They can read mathematical calculations and agree on an outside the Universe without stating it in an explanation what happened to the lost and found or they're ability of introducing the concept as a reality, which they claim it is. That such factor can go outside the Universe and leave the Universe by causing a Houdini vanishing act of never –to-be-repeated-again status. Science would have us to believe this antimatter went into hiding in a manner that is out of the Universe. They applaud this thoughtless presumption while fully knowing that at the time they do this acknowledging that there is no other place for anything wishing for a place to be within then having to be in another place other than inside the Universe. If it was ever anywhere it still is within the Universe merely because there is no other place to go than to be inside and part of the known Universe! There cannot be some factor and then misplace it as if a valid factor calculate the value can prove the disappearance and by disappearing it no longer is. If it was in the Universe it must still be in the Universe somewhere. Then we better start looking for it.

Another big issue is that what ever the Big Bang produced must be in equal terms everywhere. The Big Bang was a process that had the Universe act as a high-speed cocktail mixer of no repeating ever again. Whatever the Big Bang was of all that it was, the most it was in the beginning was that it was one massive mixer mixing everything in it at the speed of light. With all the mixing time and time to mix there was going on with nothing better to do than mix and match the mixing was done thoroughly. That we can count on. The relevancies might change slightly and balances may change favouring opposing ends…yes and known appearances did change…yes. But in the end all the factors must always be present everywhere through out the Universe. By this lacking of a fundamental explanation about what antimatter will look like when found Mainstream is incredibly poorly judged by scientific standards. Those mathematicians calculating physics suggest that science should take antimatter as a cosmic fact and then in disregard of other realities they dispose the truth by discarding its properties onto the unknown. That hardly suggests plausible science by any one's admitting.

By that Educated Scientists of High Standings are discarding even more of the old fashioned basic elementary science taught as science principles to children in schools in science classes through out the world. One thing surer than any other fact is that matter in whatever form consists of the purist energy there ever can be. In the cosmos is, was and will be all the material there can ever possibly be. Our concepts we put forwards can be faulty but nature cannot ever be at fault. Our arrangement of our ideas can be at fault, but we cannot pull a vanishing act on certain cosmic products and in doing that then dismiss the existing of such a factor or factors, which we then claim, have vanished in the further developed Universe. Our concepts of what they became may be at fault and by changing some basic principals such changing may produce a better understanding about what we think we read into mathematics. Mathematics is purely a language and mathematicians are purely translators.

Mathematicians translate from the language they read to the verbal equivalent they speak and as in all translations made certain concept may become misinterpreted. The terminology used to explain this is "lost in translation" Mathematicians must see what there is in the translation and try to incorporate what there is available in the cosmos to what the Mathematician sees in his mathematical calculations. The Universe was full of heat and it was full of material but it was not full of free space. If that is the case then where did the heat come from and where did the heat go? Hiroshima and Nagasaki taught us many things about the horror of human nature but most of all it taught us that material is heat secured in atoms and atoms are heat tightly wrapped in a cocoon, which we named the atom. Heat in any form cannot have anti in another form. The package holding heat wrapped can unwrap as it does with nuclear atomic demise. But the anti to heat is cold and cold is space.

The undeniable fact about the Big Bang theory is the accepting of a growing state in which the entire cosmos seems to be in. With all the expansion that went on we came to the point where we now are at and in such growth all aspects in the Universe must grow in relation to quantifiable progress in all different aspects, which takes us to that which is seen and that is unseen and which came along as products in the Universe where everything took everything on a growing spruce by unveiling space. That is where we now are. Such expansion include all there is including everything and not just with outer space growing. The dynamics of outer space alone cannot grow by leaving the growth of material behind. Should we wish to see where we came from we have to reduce that which we now see in our surroundings to apply to the measures that once applied in all aspects of the cosmos. Mainstream physics is over pronouncing the growth of space and with that suppresses the part matter must play in such growth by simply ignoring the issue. That Is the reason why they prefer to ignore the evidence that material is growing notwithstanding that material is growing or that their disbelief about the matter of material growing do not change that material is growing in any case. Because they cannot find any reason why material should grow they refuse to admit that material does grow. This is hiding from the truth by hiding the truth. If space grows and the Universe is getting bigger then all space grows to allow the Universe to get bigger. That includes matter and space not in matter.

Space can only grow if materials that also hold space also grow within the space that is growing with the growing space. It means that stars get bigger by the cosmos growing from the Big Bang onwards and outwards to the moment in which we are at the present. But if stars grow then the atoms forming the stars are doing all the growing as they secure more space within the space they claim. If Hubble saw space grow, the growth of space must include the growth of space holding material as well. In studying the Hubble's expanding theory we come across evidence that makes it clear that all material expand in a manner as if the expanding comes from the centre of each and all particles within the expanding space and the expanding grows outwards from every particle centre. It is using every star centre to grow from in all directions proportionally in all directions evenly. This leads one to believe that gravity is this securing of space in the material just as Kepler showed it to the world. It proves a connection with deliberate implications coming from every as well as in every specific centre. It proves that the centre $k^0 = a^3 / T^2 k$. It becomes apparent if and when separating Kepler from what Newton thought about the work of Kepler which Newton accepted as being inferior and all incorrect.

To find our birth we have to take back all growth that brought development in the mean while but the only way that that can be done is by man drawing the cosmos down to what man may perceive which forms mans ability in understanding. That is making the Universe small and as man grows man allow the Universe also to grow in relation and corresponding the man's ability to comprehend. We see the cosmos as a circle and we accept the circle because the circle is what gravity implement when the choice of form is coming from material that has all options to freely choose from. By taking the circle back one will follow or said even better we will trace the rout of the cosmos to where it then started.

All stars are many circles in many dimensions, which form when all circles join into what we call a sphere, but that leaves us only with the circles in the plural. Taking the cosmos back can only lead to one point and that Kepler told us we will find singularity $a^3=T^2k$ which is $k^0 = a^3/T^2$ k. We can only reach $k^0 = a^3/T^2k$ if we repeat $1/k = T^2/a^3$ in a continuing manner indefinitely. When one does the effort of reading this correctly, it says that when distance k brakes from singularity $1=k^0$ that is then $(k^0 =1) /k = T^2/a^3$ where the space a^3 produced a time T^2 equal to singularity k^0 and singularity k^0 is equal to eternity which was where all was equal to a never changing cosmos that was holding the single form into one dimensional space that included all the filled and vacant material filling in from all sides.

This is one way of looking at the issue and by doing that I am about to prove that singularity is Π. I am about to prove that not only is the planets adhering to the Titius Bode rule of seven over ten and ten over seven in relation to the Roche limit but that the Roche limit explains the very, very first instant the Universe experienced outside eternity. The atoms relates to space in the very same manner of seven singularity positions to ten points and from this motion of material interacting with space is securing material on the inside as well as on the outside. By that motion gravity comes about finding the value of Π^2. Gravity uses the relation of the Titius Bode seven on ten and ten on seven as well as the Roche factor to form gravity and gravity is always Π^2. This I see by reading Kepler's work as Kepler produced the work and introduced the work as $a^3 = T^2 k$. With this formula $k^0 =a^3/ (T^2k)$ must also be true because $a^3 = T^2 k$ is a relevancy that has to be in relation to singularity and therefore singularity must be $k^0 = 1$. Where will we find $k^0 = 1$?

All motion brings about results as the motion eventually ends in spin. Even our linear motion travelling along the surface of the Earth by sea or land seems to us as going straight but it is eventually following a circle around an axis. There are as many axes as there is always an axis. The axis provides a partition between the rotating directions that the spin of the material is securing at the location of where the axis will follow. The spin will have motion and the spin will have direction although the axis will forever instantly change the direction of the spin continuously to fit the linear part of the spin. By going straight the directional change singularity used because singularity is what it is, is continuing to eventually become the circle motion. As the direction will forever change the linear then will forever remain steady due to the eternal changing of the direction.

The linear remains linear because the linear redirect its intentional direction because of the rotational change that the linear motion always end up in doing. The line forms an eventual circle   because   the   linear line must constantly entertain the centre.

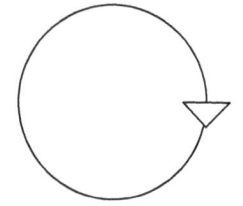

The line becomes bended

That bending repeats the action to go on continuously bending

To become a full circle in the end

Our gravitational falling to the Earth is a result of a circle going straight and forcing us straight down to an everlasting directional alternating circle we have as we spin with the Earth as we spin around the sun. As we fall straight down we change direction while we are falling straight down because that point we are heading t what we are falling to is changing too. But from at the centre of the axis everything seems neutral. The axis does not spin at all because the axis brings about spinning motion changing eternally. That is in nature and not with man made motion.

Because the axis provides and demand direction changes to secure everything in motion around such a centre to such a centre, such a point forming the axis is beyond dimension. It has no side as it has no space and it has no motion. It cannot be detected because it does not contribute to any space the Universe has but it can be located because it does contribute to all the forming of space the Universe has. Without ever moving and because it never moves, the centre forces rotation by being in the centre as that centre is also commanding from the centre. That point allows motion to apply where such motion act as the partitioning between objects. When the spin comes about say in a child toy such as a top, the top gains independence producing (as long as the energy will last) an independent motion in spin but when the motion dies the independence is lost. Previously I mentioned that all circles in the plural forms a sphere by duplication but never repeating opposing controlling points connecting to a joint circle that confirms all possibilities and re-ensures all possibilities. In the final analyses there is one centre on will reach in reducing every radii.

We accept that the time it takes a planet to move between two points is time T^2. Having that space a^3 in relation to the time T^2 is space-time a^3 / T^2 and that is precisely how Kepler expressed his findings $k = a^3 / T^2$. This indicates space-time that is growing through the extending of k. While it does prove the Hubble shift it underlines that that is not what the gravity which we experience, because $k^{-1} = T^2 / a^3$ (Newtonian gravity) dominates by contraction where the gravity permitting expanding $k = a^3 / T^2$ is not inclined to absolutely favour contraction. Newton's gravity is totally about a decline in k. But what Kepler shows that in outer space through motion of space performing space-time But our gravity does not exclude the implications of growth because $k^{-1} = T^2 / a^3$ (Newtonian gravity) allows material growth by extending.

The gravity we feel that is dominating us which is also that which Newton saw $k^{-1} = T^2 / a^3$ (Newtonian gravity) cannot realistically accommodate growth in the Universe. This should therefore remind us living a life of splendour on Earth that we must remember that we are part of the Earth and not part of the cosmos. We may find some ability to reach outer space and remain there for a very short period but then we have to return to the Earth. The returning part is compulsory and that we must accept as we accept breathing. There are many suggestions of how we can achieve the ability to distinguish mans superiority by extensive and elaborate travelling through out the entire Universe vindicating our millions of years of being confined by the dooming gravity that the Earth grips us with and committing us to our revenge by knocking off our shackles as we cross once more yet another barrier similar to that of Columbus but infinitely more, wider and holding unlimited vastness just to secure our seemingly unstoppable ability to travel through outer space at the speed of light. Here comes the shocking part: Those that cherish the hope such inspirational thoughts may bring, those thoughts as inspiring as they may sound are no more than blatant useless daydreaming that is at best and at the worst only promoting wishful thinking.

We have as much a chance of achieving that dream of visiting the next star let alone the next Galactica as we have of never aging, never going sick or never dying. Those thoughts belong to the mindless thinking pattern one will find in the muttering atheist, which is bent on proving the improvable by reasoning idiotically. The atheist practises a religion tempting them to think that if there is no God, they can take the role of God and be God. To do that they have to remove all barriers that divide the sane from the mentally incapacitated. Travelling through outer space on the breeze of a light bulb is just is not possible to do. We are born on Earth and we are part of the Earth. Only through our attachment with Earth do we become part of as well as involved in the cosmos but that is strictly because of our surety we find with the Earth and we are secured because that is the Earth which is comforting our needs like a good caring mother should. We are not naturally part of any location other than the Earth and any visiting of other cosmic locations is artificial

There is no doubt that such visits will be very short lived and even such a possibility is yet to be proven because from what information I can gather there will be dire consequences to follow which are to be avoided if man adheres to sanity instead of manic madness by promoting such attempts of visiting other locations. The Earth represents us in the cosmos and represents us in the cosmos on our behalf. The cosmos does not know life and the relation the cosmos has is not the relation we have with either the cosmos or the Earth. That relation the Earth represents us in the cosmos is that gravity, which Kepler introduced while Newton saw only how the Earth jealously hold us captured by applying the gravity that Newton saw. This far man could afford speculating with his dreams because the part of science that Kepler covered was up to now just a blind spot to science. I uncovered that blind spot with the aid of Kepler.

Let's now proceed by using this information as we chase down gravity and find what more there possibly can be which it could hide. Gravity is space moving in a circle holding that space that is moving towards a centre in relation to motion. The space is identified by another space moving in the opposite direction. Between the two there is distinguishing differences and what is in space at a distance, which cannot sustain the required motion needed to maintain the gravity that is the separating second space factor that is giving independence through motion. The motion is completely different and totally harmonised holding equality by differentiating motion. The differentiation provides the equal sustained ratio in motion. If such a ratio in velocity comparison cannot be sustained the space removes as it shortens **k.**

Maintaining the distance of **k** from moment to moment is that requirement needed to keep velocity equilibrium sustained and velocity in ratio becomes the product and the result of gravity where that is prescribing the applying conditions forming equilibrium. Only when such conditions are broken by their inability to sustain the harmonised velocity ratio does space fall away and particles come crashing down to the Earth. Otherwise such conditions are maintained and an orbit comes about. But this falling comes from a lack of motion and not tucking each other's sleeves or pulling each other around. Performing a little science experiment such as the Coanda effect disproves the grabbing on theory. Gravity is about matter concentrating the heat in space through the spin of the proton spinning and reducing space. Such motion establishes an elected centre that houses gravity. The space holding the protons secure forms a demand on space flowing to replace space by filling from "outside sources" in order to replenish the point of space reducing.

The flow or motion comes about as a result of a need to supply space with more space as the proton diminishes space at the centre by killing of space as it nullifies the motion in the centre. There then is a vacancy forming as there in the centre is no space because there can be no motion to the space in the very centre. Because there is no space with motion that specific single dimensional spot has no part in the cosmos we know. Moving towards the centre there is a re-supplying of space equal to the number of protons, which brings on the reducing of space and accelerating movement of the space between the point of demise and the point of replenishing.

This is one part of a group effort where all factors forming the group work together to provide the required gravity. This part Newton saw not. Being the master that formulised the existing laws on motion he had to detect the consequences of motion if he carefully studied Kepler's formula. He would not have brought in the idea of a force but would rather have recognised it as a natural flow of space bringing about the duration of time. Newton did admit he had no idea what gravity is and declared that gravity is a force. On that point I disagree and my disagreeing is not on the subject of gravity being or not being. I am emphasizing my disagreeing on the force aspect because every person on Earth associates the word gravity with the word force and confuse the two in concept. Life is a force but gravity is a natural and normal flow from the start of the Universe to finally bring conclusion to the Universe. Gravity is a natural motion of space a^3 in space a^3 and in that there is no force to be found.

One does require a force to resist the flow …yes but that resisting of the flow then becomes the force, which counteracts the flow. His view about declaring the presence of an unexplainable force brought about much rejection from his fellow academics at the time because it enlisted a vision that Physics were moving back to the dark ages at the time. No one can blame the others about the direction they saw science move because admitting to a force without any ability to explain what brings such a force about constitutes much to the powers pagan gods, witches and other undesirable powers had at the time. However Newton did conclude that there is strength in the centre of a sphere that produces the strength within the sphere. I then decided to investigate his remark.

We all accept that the Universe uses the only true cosmic form there is, as an overall all-containing form we call a sphere. The sphere is that form, which the Universe has to be in to form the Universe and naturally the concept that immediately enters everyone's mind is thinking about it, would be and most probably is a sphere. Everyone accept the universe as a whole will be the sphere...but why would the sphere form. If there was any one in the past that stopped for a minute to think about this question that philosopher then never stopped for that minute to write down as to convey his conclusion to the following generations. I have heard intellectuals explain it by telling students the form is used by the Universe because it is the strongest form there is, but that carries the same value in definition than to say the Universe uses the sphere because the sphere is round. The original question then still remains unanswered as being totally gone unanswered because the question still stands. Why is the sphere round and why is the sphere the strongest form one may find? So declaring the sphere as the strongest form leaves the question just as unanswered as before.

What will be the reason why the original form that we devote to the Universe will take on a sphere as natural form? Yes... I have heard in Engendering that the sphere holds any point and every possible point pushed from the outside of the sphere secured by every other point the sphere has from the inside of the sphere which then is forming the sphere but that statement is as precise as it says a woman gave birth to me and not that my Mother specifically gave birth to me. It says I can be anyone's child instead of that I am specifically that persons child. It still does not reach the answer that will stop all other questions about the question. Apparently our imagination grabs the sphere as the only form of choice and that is as correct as it is true...but why...this is apparent coming from nature as natures choice to form when material is not pre-cast to have any specific form. In such an event the gravity in that space take on by cosmic pre-cast shape the sphere as form...it is because gravity chooses the smallest space to hold the strongest force. Such a point will also establish a line we call an axis. By reducing the radius there must come a point where the ring that is in decline from such reducing is infinitely small, where it can reduce no more, where it reached its ultra limit, but at that point cannot be zero, because the point is there for all to realise but nobody to see. From the point in the centre that is no point in he actual Universe there are in one space forming a unit two points separating the unit by holding relevance and without two points there are no point.

When looking at what Kepler brought into science we find a^3 being equal to T^2 by the allocating of k. The mention of a^3 is referring to the space filling the space that is the space in at the very end of the point rotating where that point is indicating to the forming of a circle T^2. But a^3 also indicates a separate a^3 that pinpoints the allocated position of the space designated to have the smaller a^3 point out the precise a^3 that the smaller a^3 is claiming as a unit and that became the product of the motion identifying a^3 as a separate unit sharing one larger a^3 and one smaller a^3 of what all is brought about by a field invested to form the gravity. There is forever a larger space a^3 that holds a smaller space a^3 in relevance to the motion coming about in the form of T^2 and k. Then the relation between a^3 and the centre part of the larger a^3 there is a most relevant point being k^0.

Considering the manner in which the expression of Kepler's formula read one may correctly be of the opinion that a^3 is in context with the broad space that covers all of the space indicated by the length of the radius which is symbolised by k from the centre k^0 to the point indicating the immediate border of the space k. Yes that presumption is very true but also true is the fact that if there was one point reserving the position for the smaller point a^3 that held a separate and independent space a^3 within the larger space a^3 which would without the smaller space a^3 not be identifiable as forming the unit a^3. If there was no such a smaller space a^3 within the larger space a^3 producing the outer limit to the larger space a^3 the larger space a^3 would have no independent relevancy in the overall totality that will distinguish such a space a^3 and to establish the containing as well as reserving position it holds. The larger space a^3 is there because of the motion of the smaller space a^3, which validates the larger space a^3 to be a factor worth of being calculated.

Only by the motion of the motion of the smaller space a^3 can the larger space a^3 claim validation and on the other end also apply independence because as I shall show later on that the motion of the larger space a^3 validate the counter motion of the smaller space a^3. The smaller space a^3 cannot be in motion if the larger space a^3 do not contribute to a larger motion of space a^3 contradicting the smaller motion by direction where both accommodate each other by motion relevancies bringing individuality without bringing independence about. Kepler said $a^3 = T^2 k$ therefore if there is space a^3 such space a^3 has to be in motion $T^2 k$ to allow space a^3 to be and have the other space within. Therefore by referring to a^3 one establish a relation of both in the context because not one of the two would be if not for the presence of the other a^3. When referring to a^3 one refer to the larger a^3 which is containing the smaller a^3 as much as one distinguishes the position of the smaller a^3 proclaiming the area of dominance of the larger a^3 in which the smaller a^3 takes up residence in space a^3.

Kepler's formula first drew my attention to singularity in the way he formulated his formula. The most important part of his formula is not visible from the outside or from the onset of investigating and one must look for that most dynamic part covered by the mysterious coming from way within. Kepler shows us that the truth is found in the darkness and not in the light. At a point where Einstein said gravity begins we will locate Kepler's gravity beginning because space (or as Einstein referred to it) the Universe goes flat. If $a^3 = T^2 k$ is a fact then there must be a starting point where k starts because there is a point where k ends. This then will change relevancies and will mathematically equate from $a^3 = T^2 k$ to $a^3 / T^2 k = 1$ and one can be any number or symbol to the power of zero. Mind you not to the value of zero but to the power of zero $a^3 / T^2 k = 1 = k^0$. That means one has to reduce k to a point where k becomes k^0, then in accordance with Kepler's advice I proceeded...Kepler said that from the smallest space within space a^3 there is the line k, which is connecting in a motion covering the spaces $k T^2$.

The space indicated by and that is a part of space a^3 in question wills run as the space-time unit $a^3 / k T^2$. That is where gravity will form being identifiable as a unit at a specific centre from k^0. Gravity lurking in the centre at the point k starts the line k where the line k holds space-time a^3 / T^2 secure and in form. That has to form singularity and singularity can be whatever there is a wish for as long as the wish is to the power of zero. (Singularity) $k^0 = a^3 / k T^2$ which reads that in space-time has three sides on the one side and are opposing the first side by three other sides. If Kepler said mathematically the smallest distance between structures could at the least be $k^0 = a^3 / k T^2$ and we all know that $k^0 = 1$ then it should be some one's duty to find that point.

One must then start by accepting that Kepler also stated there cannot be nothing or zero in the cosmos since the smallest distance between two structures is k^0 which is one and not zero. I wish to introduce an argument by disposing another Academics method in his disposing of my work. Some academic found a way through which that particular Academic was able to dismiss my arguments on the grounds that the solar system was not formed at the Big Bang period or that is the information that Mainstream science is promoting.

That was a loophole he suggested because he was unable or inferior or plainly just to lazy to interpret my work that I laid before him. I am not for one minute fooled by his passiveness because that is very typical of the New South Africa everyone outside South Africa helped to create. And in addition I really can't think that he thought me to be stupid enough to be discouraged by taking his arguments seriously. In order to circumvent such a loophole I shall begin my following argument by stating that the solar system, which I am referring to, is a hypothetic one. Notwithstanding that I know my argument is solid and serious as such about all aspects in the rest of the entire argument which include all other possible aspects, the following is the one part I use that as a part of my argument, this part I now identify as a part remains the only hypothetical possible fact in the entire argument that has a possible hypothetical truth as it stands.

There are those who avoid admitting to inconsistencies by arguing that my argument about growth of material in the cosmos throughout its entirety is invalid because the solar system was not in place at the time the Big Bang was in place. Please then keep in mind while reading the argument that I would like to point to the fact that my following referring to the solar system is actually referring to a similar solar system that is somewhere else and is now a part of a galactica we do not know about. That is where the dissimilarity ends. In all other aspects our solar system and the one I suggest is identical in every possible aspect I bring this in to disqualify any academic loophole that may come about from an argument about the solar system coming into place at a later stage of the cosmic developing and therefore I exclude any chance of using the counter argument that the Solar system became an eventuality long after the Big Bang was about to happen.

The novelty in the forming of the solar system in the argument I am about to present is no longer an issue because I am referring to a solar system in another Galactica being precisely the same as the one we use with the exception that that solar system was around at the time of the Big Bang. I therefore hope that there is no more ambiguous loopholes that any one may use to unfairly dismiss my point of view when the validity of dismissing my point is as simple as raising such an counter argument as the one mentioned. That counter argument of the solar system not forming a part of the Big Bang does not apply.

To avoid such a loophole again we now use a hypothetical but real solar system in space, which formed when the Big Bang took place. To them we now present a solar system that is identical to the one we know and is a precise duplication thereof. Again I say to those the argument now represents a precise duplication of our solar system and was in place ever since the Big Bang. That means with the solar systems being apart in millions of kilometres at the present time there was a time when the planets and the sun were apart by the same measure but only using kilometres instead of kilometres by the millions. The Big Bang shows a growth in space. If that is the case then there was a time when the Earth was 149 kilometres away from the sun instead of the current 149 million kilometres, which it is at present. We can reduce the distance further to fit into a billionth of a meter but I hope my statement drives the point home. But that space we reduce also has to include the expansion in size of cosmic objects because reversing the expansion shows that any argument not expanding the orbiting structures is most silly.

Then there must have been a time when all the planets were between fifty-nine and five hundred and ninety metres away from the sun in comparison to the millions of kilometres we have today. If there was no expansion of the orbiting structures in the diameter size they have then how did the planets being the size they are at present, which then was also the size they must have been back then, fit into such a space that small where the where that small space was keeping such giants apart. That is notwithstanding the fact that they still are with all the gravity they presently have but at the time was being apart only by the measure of 149 kilometres as is the case of the Earth?

Material therefore too must be part of the growing in space. This line of arguing I suppose is much below the Academics pursuit of matters but since I am much lesser in mental standards of developing than they are, such reasoning prompted me to go on some investigating journey. Light journeys through out the cosmos and it will be sensible to follow lights travel in reverse to see where that takes us.

With objects being apart at some distances and light flowing in straight lines between them it must take light the size of a straight line to travel between cosmic objects. While I disprove this statement in future arguments I wish to stick with the officially accepted but as such a very simplistic concept for the moment. The distance the light has to cover depends on the radius there is between the objects and that forms the total distance forming the Universe. That put the Universe then in relation to lines forming differences about structures that is claiming independent space and space setting objects apart as such distances will be the radius standing between those objects. In reality that is what the Universe comes down too.

The objects are circles by dimensions that are the dimension. With Bang coming from cramped for correct path to steps of the Bang is to between the such a line will realised I had had to find the lines end when I which will then be a point will show me cosmos started. The material structures all sphere is a lot of circles the other candidates. another name but to the beginning of where the sphere in its

and establish a link with the centre. The Coanda affect is about space in motion acknowledging a centre formed by the motion of the space in motion. It supports a centre formed in acknowledgement of such a centre formed at random.

dimensions and the space is also crossed by lines travelling through light being a line and the Big a situation that was a lot more space than at present, the follow if I wish to trace the cosmos back to the Big reduce the straight line structures and find where no longer be a line. I to begin at a point where I point where any and all reduce any and all lines the same measure where such where the line forming the same procedure will apply to the being in a sphere form in our Universe. A forming a unit but not repeating the space claimed by Such a circle also applies a straight line only known by still serves the same purpose. Reducing the line will lead us time. The reducing of the line will once more represent the point role as a multi circle will begin.

Gravity forms as the earth or any other cosmic body rotates by 7 $^\circ$. By diverting the straight-line movement by 7° a contraction forms in a circle. In my books I prove how this then brings about the value of Π by implementing the law of Pythagoras and gravity is the law of Pythagoras.

The reclining of space by redirecting the direction of travel from straight ahead to 7° reclines the space in a steady and sturdy flow. It is the space reclining or contracting and the space contracts albeit filled by solid material or empty of solid material. This is the reason why all things fall equally. It is the space moving down with or without holding material and the space has the same density in relation to the solid cosmic structure rotating.

In the centre of the rotating body singularity forms to the value of Π^0 and this extends to the curve forming Π in terms of the curve of the rotating object. This then forms part of gravity moving forming singularity going square Π^2 and this form the relevancy

The extending of singularity goes from Π^0 to $5\Pi^0$.

Coming in from space while having no mass the coordinates change as the value of Π forms a relation to the 7°

When an object comes in from the atmosphere or flies the sky the object associate with the turn giving it a value of the rotation in line with the axis which puts a value of 21.991 / 7 on Π or on gravity. This is gravity and this is how gravity functions. It has nothing to do with mass in any way, shape or form and there is no factor such as mass in the entirety of the cosmos apart from being in the imagination of the Newtonian conspirators calling them physicists. All gravity is determined by movement and not by mass. The entirety thought of as the Universe is linked by singularity because in the centre of all thing turning we have singularity and because singularity is $1^0 = 1^1$ that would be the reason we can see the entire Universe through our eye lenses. But since $1^0 = 1^1$ what there is as far as the smallest and as far as the biggest space is just imagination and the truth lies beyond our scope and beyond out vision and beyond the Universe.

This should lead any person to investigate a centre that forms because evidently there is a centre but that centre comes about by motion of rotating around a fixed point serving all points in motion. That leads us to the centre of everything in rotation because everything in the Universe is in rotation. As Kepler said $k^0 = a^3 / k\, T^2$ **and we have to find** k^0

Locating and finding the presence of singularity

$k^0 = a^3 / T^2 k$ **states that whatever is, is also spinning in order to be present.**

What is in the Universe is spinning. In the **precise middle** of all **objects in rotation** is a precise centre dividing the object in sectors that will **start the spinning initiation** from that centre point. Thus, the spinning object **will have a middle point**, a very specific **centre point that does not spin** and only holds Π as a specific value because no radius can apply. But also the one value such a line **cannot have is zero** because the line **is there and holds contact** to the rest of the material bringing about that **zero does not start any** line and therefore the **value of the line must be infinite**, just as described in **accordance** and by **the definition of singularity**

As I am introducing a very new idea, I whish to explain in better detail what I try to convey.

While the toy top is spinning one will find singularity by moving the rotating line or radius progressively to the middle by reducing the length the line have from the edge to the middle. At one point all further reducing must end but the ending cannot include zero or nothing because the rest of the line still attaché the rest of the top.

That point albeit hypothetical, is also as much a reality none the less and is placed where that point **must be standing still** because every line **running from that point** in **opposing directions** are also **in opposing directional spin** the other or opposing side.

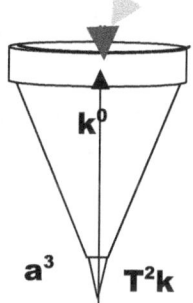

k^0

a^3 $T^2 k$

k^0

k^0

> As the rotating direction moves inwards, the rings will become smaller and smaller.

In considering the spinning motion in the fraction of time in the detailed instant every aspect of rotation will turn in every instant of change in time. Although the points had the same characteristics only one instant before, they oppose the characteristics it had just before and just after the very instant in which they are and to which they relate by similar points also in rotation. The fact of the graph proves my point in quarterly opposing dimensions and values,

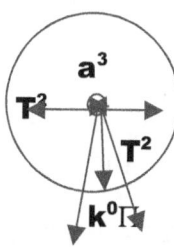

a^3

T^2

T^2

$k^0 \Pi$

$k \Pi^0 k$

In dimensional terms, which I explain later on the value of **2k** relates to T^2. That relation extends to the next value where T^2 relates to **k** , which relates to T^2. The first space in the circle will then be T^2 **k**. From the centre being in infinity one can realise by applying mental power the single dimension factor not seen but present all the same. Extending that into the 3D comes six **k** and any one of the six will further extend to form a seventh point as T^2 All this is a multiplying of $k^0 = a^3 / (T^2 k) = 7$

Water flowing will release from the normal line that gravity enforces

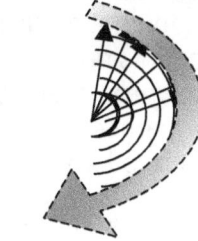

All this was missed by science ever since Newton changed Kepler's work because Newton failed to read into Kepler's work correctly from the information he (Newton) basically discarded due to the arrogance on the part of Newton. He (Newton) concluded every factor in the whole formula of Kepler

represent motion instead of the true representation it carries being space-time.

He failed to study Kepler's work even though Kepler arrived at this work only when he (Kepler) received the work directly from the cosmos as Kepler was studying the cosmos. In ignorance on the part of Newton driven by arrogance he failed to notice that Kepler's work introduce for the very first time space-time as a formulised conclusion and being space-time Kepler's formula has no need of adding Π and having the formula changed from Kepler's vision of $\mathbf{a^3 = T^2\ k}$ to $4\ \pi^2\ a^3\ /\ T^2 = G\ (m + m_p)$.

Newton tried and failed to marry Kepler's formula with an earlier concept of him (Newton) which he (Newton) received by vision when he introduced the formula $r^2\ /\ (M \times m)$, which was his first vision of what gravity was. With that he formalised gravity but then he overstated the grandeur of his first vision by matching his concept to the Creation at large which was introduced by the work of Kepler. In that attempt to link all creation by what he found to be condensed in the name of gravity he changed his first formula from $r^2\ /\ (M \times m)$ to $F = G\ (M_s \times m_p)\ /\ r^2$ which Newton then afterwards introduced accompanying his term as gravity. This was one of the most unfortunate and corrupt misleading in science that ever took place.

Consider the seriousness of misguiding that came from the case of the Piltdown "Ape man" Hoax and by comparing the influence such diversion inspired the Piltdown "Ape man" incident becomes an innocent little party prank compared to the implications that sprouted from the Newton diverting of science. But it is not only Newton that is charged with the deliberate misleading in science because Newtonians are aware of the so many missing answers to questions Newtonians never dared to ask in their attempt to hide all the shared blame.

By merely diagnosing gravity being naturally pulling from the centre is rather avoiding the question with simplicity because the question arising from this answer is where then is the centre of the universe to where the final pulling is heading? By using the gravity formula $F = G\ (M_s \times m_p)\ r^2$ the centre is the vital issue directing as well as controlling every aspect involved in the formula. When this formula is mathematically translated to English the formula says there is a pulling of structures $M_s \times m_p$ towards a centre r from both sides r^2. There is the centre r^2 that is the sun. There is the centre r^2 that is the Milky Way. But that is not where the Universe stops.

The Milky Way centre is not an individual end-of-the-finality-of-the-Universe in terms of coming from as we are going too. Mainstream Science declared that all the pulling is toward a centre r^2. What placed r where and what was r be to achieve that much pulling power? We see that the Universe is somehow always coherent and disciplined because the rotating of structures indicate the presence of gravity.

The Newtonian formula insists on a centre pulling particles. In the formula $F = G\ (M_s \times m_p)\ r^2$ such a centre is most important. Such a centre is as demanding as it is commanding and yet, Mainstream Science never came to pinpoint the centre of the Universe. Kepler on the other hand showed the precise location of the centre of the Universe. The pulling of all mass in the Universe must be towards a centre and because of Newton's introducing of a Gravitational Constant (G) in the Universe; this then demands a centre to form G.

Such a factor with gravitational powers must point all gravity forces towards one centre since gravity is undeniably located in a centre of all objects. When the realising of this all out important centre arrived and Newtonians became able in recognising this factor Mainstream Science should have then at the time have been working towards identifying some centre before proclaiming the serious implications arriving from the presence of such a centre. Never once was there too this tome some launching of an all out search to locate this centre.

There was a silly attempt designated to Einstein because of his superior mathematical abilities to calculate the entire mass of the entire Universe but what prominence would the mass have in pulling without a precise indication of a specific location where the pulling of the entire mass is heading. Without the centre all other factors loose their validity. Where then will such a centre of the Universe be? This is what makes cosmology the shambles it is. Kepler gave us the answer centuries ago but no one ever tried to take notice of what Kepler said without Newton's interfering. Kepler gave us the ability to see what lies beyond the limitations of the visual as well as the very obvious.

$$k = a^3/T^2$$

Kepler said $a^3 = T^2k$ but that could also be $k = a^3/T^2$

When translating Kepler's mathematical expression into a verbally spoken form of communication such as English we can see what Kepler said also read as $k = a^3/T^2$ where k is one point from a centre point that is space a^3 relating to time T^2. From a centre comes space-time

Others like Newton and Einstein came much after Kepler and coined the phrases but Kepler formulated the concepts of gravity as well as space-time. They (Newton, Einstein and many others) named Kepler's innovations. That is very clear but only on the condition that Kepler is read correctly and Newton gossip about what Kepler is saying is ignored. What Kepler said in mathematics all the brilliant Mathematicians through so many centuries were unable to read although the coded language was written in mathematics and as it is their field of supreme speciality they have captured as theirs, therefore they should have the ability that would enable them to decode what they are the masters of which they claim to be able to decipher. It is the mathematics, which the cosmos use to allow the cosmos such communicating with humans! The cosmos said that all space stands excluded form all other space which then forms an all including unit.

That is why the sun is so cold and outer space is so hot (and no… calling the sun cold and outer space hot is not a mistake on my part and neither is it due to a printing error but the sun is as cold as it gets and the outer space is as hot as it get. We'll get to that one later on…). The two share a cosmos in which they are both apart while forming one unit. The separating part in the unit is motion driven by heat and the motion as well as the heat carrying the motion that sets the two apart while both are in the unit. I mentioned previously that there is an undeniable connection between heat and gravity. Let us do some investigating and try to establish answers. is read correctly and Newton gossip about what Kepler is saying is ignored. What Kepler said in mathematics all the brilliant Mathematicians through so many centuries were unable to read although the coded language was written in mathematics and as it is their field of supreme speciality they have captured as theirs, therefore they should have the ability that would enable them to decode what they are the masters of which they claim to be able to decipher.

It is the mathematics, which the cosmos use to allow the cosmos such communicating with humans! The cosmos said that all space stands excluded form all other space which then forms an all including unit. That is why the sun is so cold and outer space is so hot (and no… calling the sun cold and outer space hot is not a mistake on my part and neither is it due to a printing error but the sun is as cold as it gets and the outer space is as hot as it get. We'll get to that one later on…). The two share a cosmos in which they are both apart while forming one unit. The separating part in the unit is motion driven by heat and the motion as well as the heat carrying the motion that sets the two apart while both are in the unit. I mentioned previously that there is an undeniable connection between heat and gravity. Let us do some investigating and try to establish answers.

From since the time that man discovered intelligence man (if he ever did) has been with the presumption that the sun is the hottest centre in the solar system. Later on in the more present time it came to someone's attention that the sun also holds the solar system in gravity. The Earth by its standard and dominating its sphere of which it can control with influence is the hottest centre in the space of its domain and it holds the moon centred to the Earth.

The gas planets are the hottest centres in the relation with the most heat and they all hold their satellites captured by a hot centre. All space structures hold in every centre there is that is confirming their independence at that point of securing independence the centralizing of the most heat it is able to concentrate and from that centre holds all material captured or controlled in the domain of what that forms the independence of the structure. I can go on and on but heat in the centre couples gravity to space-time, just like Kepler said before he was spoken for on his behalf and without his permission or his agreeing to it.

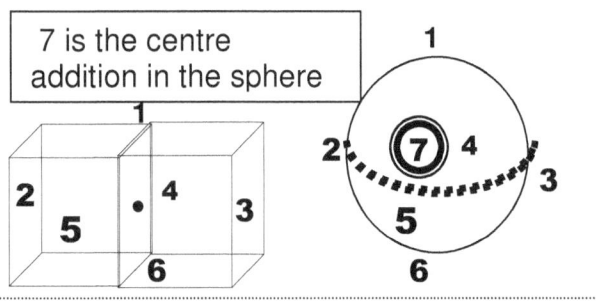

7 is the centre addition in the sphere

Kepler's formula also indicate that a sphere is within a cube that is holding a sphere

Taking the outlook from the point the sphere is holding from that centre out into space there are ten points connecting to the centre. In that are the dimensions of singularity connecting to space where five connects to space in the second dimension of singularity, and five connect in the third dimension of singularity. On the other hand does the cube show a very different characteristic, which involves only six sides (at least) connected.

$a^3 = (T^2 k)= a^{3 +2 + 1 = 6}$ with the sphere presuming the position of singularity as part the of $k^0 = 1 = $ **singularity**. Einstein proved that at the point where space reduces and such reducing reaches a point where space as a factor in the third dimension disappears into the single dimension (space going flat) gravity is overwhelming. Einstein interpreted this, as the complete Universe going flat while the Universe going flat, that can only be within singularity since singularity represents the Universe as flat as it can get.

Humans (including Einstein) interpretation of the Universe is faulty but the faulty aspect does not include the fact that the Universe is going flat but only which is the flat going Universe referred too. According to Einstein he proved that the Universe is alternating between going flat and holding space but his lack of studying Kepler lead to his spontaneous misinterpretation collected from our culture and his incorrect interpreting of what the Universe actually is. We all have a faulty perception of the Universe because not only he (Einstein) as an individual Scientist but all humans throughout has also never asked the Universe what the Universe is. Kepler did and the Universe answered using the mathematical equation $k^0 = a^3 / T^2k$, which when interpreted means singularity placing space-time is the Universe. No one ever thought about this statement in sincerity because from a Newtonian aspect it seems silly. But rethink the silliness presented by the Newtonian Universal centre and compare that thinking about what the Universe told Kepler then decide what is silly. Newton's never acquiring the effort to do a study of Kepler's work withheld him (Einstein) from reading his very own mathematic translation accurately because apart from Newton Einstein must be considered the second most important Newtonian ever. What Einstein saw was that space disappear and he then jumped to the conclusion that the space he saw in his mathematical equations was outer space referring to the space falling outside the parameters of the material occupied space secluded by dimensional borders. In the sphere placing the borders that the sphere holds there are deliberate and very distinctly placed edges or points forming a specific distance from the centre. The centre is also proven beyond any debating.

The centre of any sphere has to be at the very point where space completely falls away. It is at the point where all the points of line centres meet by the crossing the centre of their individual connection coming in to contact as a group. In that way one may assume that the lines connecting the controlling points on the other end is crossing on a centre point that all that is participating in the constructing of the sphere is democratically electing such a centre. Please note this conclusion very well because this forms the heart of the Coanda principle. That will put that position where the lines cross which in itself is centralising all space in the sphere at that point, such crossing point will become very distinct and controlling where that point forms in the single dimension and singularity is the single dimension. But Kepler also solves another riddle that truly got Newtonians unstuck. This too which I now refer, is what is referring to when they refer to the Hubble constant.

The growth we see in the Universe is an adding of space in every cycle completed by every cycle, which all the protons complete. The adding is the smallest addition that can come about in the shortest period of repeating by cycle rotation there can ever be. This growth of space-time next to singularity confirms the growth of singularity as singularity recall the space it uses to grow in the time it grows. The margin of growth will be by the extension of k in the formula $k = a^3 / T^2$. Every cycle completed in the relation to space by the initial value of k. $k = a^3 / T^2$ leaves ultimately a^1 extending as space or as Kepler chose to indicate it as k^1. But that two has to be compensated by the duration of time reducing the time aspect by the margin that the space expands. This confirms what is evident in the Hubble Constant. The further one look at time the more time seems to race because time has the invert properties we give to space.

I suspect this cosmic growth of all material is equal to the growth of a human hair or a human nail the presents as the duplication of cells because life takes command of what is made available by the Universe and then manipulate space-time to claim such growth by taking charge of the opportunity to use the growth to the benefit of life. But this growth constitutes multiplication from the very centre of the most inner part of the where k = infinity plus one.

Kepler thus gave us the answer about what Hubble found what was happening in the Universe centuries ago and centuries before Edwin Hubble's discovery. From Kepler's formula one can see that time and gravity is the same because as gravity weakens so does time reduce and as space expands so does the influence of gravitational reduce because gravity has less time per unit to control; more space per unit. Gravity is $T^2 = a^3 / k$ since the object cannot depart at any further distance between the centre and the object and is captured at that distance. Also gravity is $k = a^3/ T^2$ for

$k^0 = a^3 / T^2 k$

the very same reason. The circular bonding T^2 of space a^3 is enforcing an orbit T^2 to gravitationally circle around a specific centre k, which indicates the gravity $T^2 = a^3 / k$ in relation to the other gravity component $k = a^3/ T^2$, and it means T^2 is a circle of gravity and k is the straight-line distance of gravity applying motion. Still Mainstream Academics ignore my statements that gravity is space in motion and motion of space is time: precisely as Kepler said. Any area to the cube is space a^3.

If particles were that close and yet they expanded it is proof that gravity is not about particles pulling each other closer because then the Big Bang provided the ultimate opportunity to unite what there was instead of expand what there is, but I came to realise that gravity is about removing the space between the body and building on that which the body already hold as well as the space surrounding the material that serves as unoccupied space whereas material holds occupied space. A fan drawing air into the blades has the same sucking or pulling that one experience with gravity but in the case of the fan we know it is a air that is in motion and the motion extends to affect those objects placed in the line of the air flowing. Gravity comes about as space a^3 applies motion T^2 and singularity k gravity is as much part of

Fan contracting air by producing flow of space-time and not just air.

from space as the motion of space is part of gravity.

Mass is the result of applying gravity by reducing space. Protons are the only diminishing devises of space-time there are in nature but the protons remove space as it concentrates space-time to furnish material with growth of material. The more protons there are allocated to a specific space the more space the larger number of protons will diminish. Gravity is not the result of mass. The belief of mass brought about by large numbers of protons confined in little spaces is not always true. In some cases the mass produced by a large number of protons does no result in a heavy confined element since there are those with high number of active protons which should enforce a large gravity, the opposite is true as the element show as an elaborate anti gravity by being an airborne element or a gas. There are those holding protons in clusters with numbers matching heavy metals but that is categorised as gasses and gasses produce high ratios in antigravity.

Gravity is the result of the motion of the number of protons that through such motion creates the reducing of space in the space less centre. Gravity dismisses space and by doing that the stronger gravity is where more particles can have fitting into less space occupied where that reducing of the space is bringing about extensive mass increases into the volumetric occupied area confining more material into less space. Gravity is space measured over time. Gravity is the space in comparison with the time affecting the space. It is the motion of space relating to the time of motion of space. Gravity is the moving away of what fills space by extending singularity where singularity responds by bringing about space between the structure and a centre being within another and larger structure. The larger structure in consider to fill the role of the governing singularity will hold more material in less space which is then more material that is confined to less space.

The centre structure is reducing more space in the time factor between the moving structure and the centre structure. Gravity is the increase of heat occupied by the reducing of space in a spherical unit. If Newton only tried less to deny Kepler any recognition and gave Kepler more deserving accrediting about Kepler's input in the total work, he Newton would then have seen that that was what Kepler formulated gravity to be. Kepler formulated gravity as space a^3 over time T^2 in relation to a centre k. It is the space that relates to time in relation to a centre just as Kepler introduced gravity to be. Kepler said space a^3 standing is over time T^2 in motion k. $a^3 = T^2 k$ and to all those who tries to give space-time some godly appearance with mystic properties can lose the séance- like attribute they wish to connect to space-time. Every bit of space however insignificant or however demanding forms a relation with time, which is what separates the different space from one another. It is the separation coming from time differences that distinguish space from one another. .

Gravity is working on a principle of indicators pointing dimensional integration and separation of space through heat densities applying different grades of space intensity. That means the space does not mingle, but forms layers. This is unlike one would expect from the advocating by Mainstream science about the characteristics of space. By gravity acting space becomes denser and therefore space can become a liquid and as all liquids do, space then depends on specific densities being in specific positions. With the specific densities borders come about in space. It is as Kepler stated gravity to be even before Newton came up with an idea that there was such influencing going on and named the influence gravity. Gravity is $a^3 = T^2k$, which is the space a^3, that forms through the moving T^2k thereof giving the space a^3 independence as the independence comes about of speed differences which is motion in relevancy which is T^2k. It is distancing a^3 from k by applying T^2 in the surrounding space and this is done by a^3 duplicating in motion when applying T^2.

The Oxford dictionary of Astronomy defines gravitation as follows
Gravitation is the force of attraction that operates between all bodies. The size of the attraction depends on the masses of the bodies and the distance between them; gravitational force diminishes by the square of the distance apart according to the inverse square law. Gravitation is the weakest of the four fundamental forces in nature. I. Newton formulated the laws of gravitational attraction and showed that a body behaves as though all its mass were concentrated at its centre of gravity. Hence the gravitational force acts along a joining the centres of gravity of the two masses. In the general theory of relativity gravitation is interpreted as the distortion of space. Gravitational forces are significant between large masses such as stars planets and satellites and it is this force, which is responsible for holding together the major components of the Universe. However on the atomic scale the gravitational force is about 10^{40} times weaker than the force of electromagnetic attraction

Gravitation Constant
The constant that appears in Newton's law of gravitation. It is the attraction between two bodies of unit mass at unit distance apart. Its value is 6.672×10^{-11} N m^2/kg^2 when the distance is expressed in metres and the masses are in kilograms. Although it is described as a constant, in some models of the Universe G decreases with time as the Universe expands (see Brans-Dicke theory), but there is no evidence for this.

Gravitation Mass

A measure of the quantity of matter in the body. It is measured in kilograms. Mass determines the strength of the gravitational force exerted by an object.

Now even three hundred and fifty years on, science still came no closer to explaining what gravity is in contrast to the fact that Mainstream science established even more forces than the one that Newton declared at the time. If Newton only was less presumptuous about his genius and took more notice of Kepler's work he (Newton) could have seen just what Kepler said what the cosmos told Kepler mathematically about what gravity is. That effort would have saved so much misconception. But even almost four hundred years on Newtonian disciples will not recognise my personal effort to indicate too the world what Kepler said what gravity is. I have been trying to indicate this thinking by using academic channels but on grounds not related to my effort I was dismissed so many agonising times by so many academics in charge of Official policy protection.

Although it is most apparent (to me at least) that I can tell what Kepler saw and tell them that, still they the Newtonian priesthood silences me just like Newton silenced Kepler. Newtonians should have realised centuries ago that Newton and Kepler did not have the same mathematics in mind. Consider the following and then decide

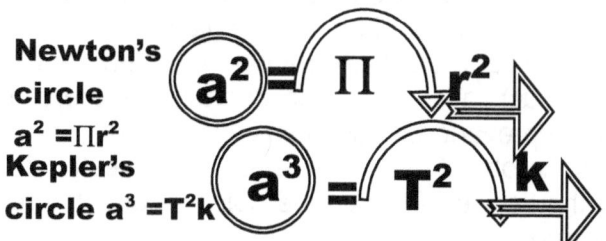

Newton's circle
$a^2 = \Pi r^2$
Kepler's circle $a^3 = T^2 k$

In the manner the two measured the circle Kepler used a different measure about the circle than Newton did. In the case of Newton the radius r goes square while the circle Π indicate the single factor. Kepler deliberately ignored the factor Π for reasons I explain elsewhere.

Kepler saw a circle because space is motion provided by singularity from a centre.
Newton whished to see a mathematical equated circle and the formula therefore was in need of revision.

Kepler intentionally stated the very opposite of Newton because the circle indicator T^2 goes square and the diameter indicator **k** remains single. That is how the cosmos relayed the given information too Kepler and that is how Kepler correctly interpreted the given information but Newton thought himself as being brilliant enough to change all that in favour of his ego. While Kepler formulated his ideas according to cosmic information Newton saw this effort of Kepler formulating the cosmic numbers Kepler measured as mathematically being incorrect. The question Newtonians failed to ask is: How can the cosmos give mathematically incorrect information about the cosmos?

Let's look at gravity while we use our common sense.

There are three factors forming gravity. In the one scenario gravity is all about the motion of one part in space applying relevancy to another motion by dismissing space through the effort of moving through the space in an a common factor. effort to depart from the centre spot where the gravity is vested as the strongest applying influence. The effort behind this is speed or if you wish I can use the term velocity but it will still mean it is the comparing of different motion each holding a different value but still requires

The object in motion is displacing the space that the object occupies in relation to the surrounding space and space being between the object and the centre. The object is dismissing space by motion as it is in motion through space. But there is another factor just as relative that is also applying motion. The applying of motion by the departing object only holds any relevance to the centre from which it is applying a distancing or an escaping attempt by increasing the distance between the centre and the escaping object.

On the side of the escapee there is another space active, trying desperately to escape by enlarging the influencing field the escaping singularity is fighting to establish. Seen from the escapees point of view in relation to what the escapee tries to bring in place what is important to the escapee is that the only motion there is the escaping object trying to escape. However much the escapee is contained the escapee is only aware of its motion it finds as it sees all motion being only progress in an continuous charge of escaping.

Just as Kepler stated gravity are three relevancies acting in opposing as well as sustaining motion where the one action works by interlinking with the other without being aware of the other. We have to recognise the relevancies applying. It is motion performing contrasting relevancies and without all three independent actions of motion contributing as one not one act of gravity is possible.

There are three relevancies applying motion in the formula Kepler left us. There is the space in which the formula presents the dimensions. The centre provides the space the dimension to move through the space as the object is trying it's utmost to depart from the centre as far and as fast as it could. It is displacing the space in which it is by it is in using a time as short as a period as it can manage. It is running of into the distance from the centre. In this motion the smaller object tries to repel from the larger object claiming to have the centre of space while it is still part of the bigger space in which the smaller space part of.

While the orbiting structure has the effort of displacing the occupied space the material hold in an effort to produce a cooling effect on the material in the occupied space it holds, there is another centre in another object that has to be much larger than the roving orbiter and as such the centre of the centre object is removing space towards such a centre in an effort to secure the centre of remaining cool and thereby prevent destruction by overheating. From the centre object there is no escaping object but only space it is harvesting in an effort to sustain singularity in the centre of the centre object is and where the strongest pont of gravity is located.

While the smaller but independent space is busy with the great escape effort by putting a distance **k** between the space in motion and the centre it fights to secure an independence from that centre. The larger space is accumulating as much space as it is contracting all space towards the centre. The independent and escaping object is moving through the space surrounding the object in this effort to put distance between the object and the centre. If it did not do that in motion it would hurry towards the centre instead of circling around the centre year after year. From the smaller objects vantage the smaller object is carrying through space displacing the space surrounding the object by motion in space. From the point the smaller object has is the space is standing still while the object is applying motion to get away from the centre. The object is applying motion by dismissing the space it is moving through.$a^3=T^2\ k$

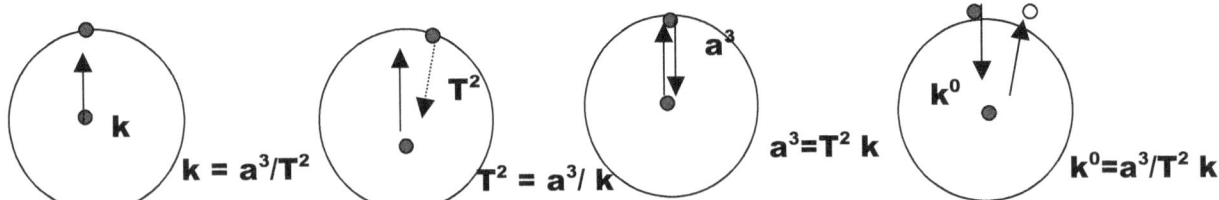

By duplicating the space of any particle sharing space within a larger cosmos structure such as an atom inside a star or a human inside the Earth there are two relations applying

The factor **k** represents as much space as a^3 as a factor represent motion. But **k** also represents as much space as it represents motion because it represents motion

This means that all space in motion is resisted by space in motion contradicting each other. In applying motion to space there is never just acting but always counteracting, which is precisely what Newton's third law on motion states. Because there is a motion in one direction there will be a counter motion performing a balance to the reaction on the motion and establishing a motion be a reactive motion. But also it proves Newton's first law on motion by proving that equilibrium existing of the action of space in the counteraction in space brings about equilibrium and being in motion that is in countermotion provides for the same effort as being in rest or in a stable state equal to being in rest after forming a circle eventually. Since there is equilibrium between the motion of expanding and the motion of contracting any additional motion adding to the "stalemate" existing will bring in Newton's second law into action in the manner Newton described.

What this formula, which Kepler introduced tells more than any other fact is that there is no space if space is not in motion and all space there is must be about motion or else there is no space. That is establishing the fact that can only be by motion and motion is what space does to move from one point in time to another point in time by the square of such motion. If there is space the space is duplicating space by projecting space from a past through a present to a future and that means the space is duplicating what was in the past towards the future through a present. But if the space and all space there is, is not duplicating by motion it does not qualify to be in the Universe. The fact of space $\mathbf{a^3}$ is about the equal $\mathbf{=}$ motion $\mathbf{T^2\,k}$ thereof.

The space we find is not just space but the space we find is motion of space as the space reduces to a centre by duplicating or by motion of repositioning in a rotating relation by doubling what there is in a repeat thereof. This motion, too which I am referring, is extremely apparent in the illustrated imaging the functioning of a Black hole where one can witness the total collapse of space toward and around a mathematical centre point forming where space and motion ends.

The containing structure represents duplication and in that it applies a norm set on conditions applied by the governing singularity of the space- time, which is supporting and maintaining the point in singularity. By containing the subordinate singularity with the ultimate goal to initiate the inevitable uniting which all singularity will do in the ultimate end of all development when all singularity will once again re-unite. But in the mean time all singularity is fighting the fight of their life to remain independent, as much as possible and by what measure it may have to produce. When the two are mostly on equal motion the gravity each produce will bring about motion of maintaining independent by creating space through duplication. When the motion equality falters the space reducing of the dominating space will relinquish space by contracting space faster than the dominated space can accomplish by using the duplication of space. When *The duplicating sector of the star...* the dominated space falls prey to the incorporating of such a smaller space into the domain of a larger space the fight for independence is carried on but the terms of conduct changes. As the dominating space adheres to the containing of all space within the domain of the larger space the lesser space will be reduced to the *The inside part is dismissing sector of the star* point it may have to relinquish independence totally. By not agreeing to such terms the lesser independent singularity continues the fight for independence by an all out resisting relinquishing all individual form and to join and unite singularity. *...Is the outer side of the star* Loosing independence will amount to relinquishing all forms of independent space being independent because of the independent motion it contributes too the unit of space –time within the unit of space–time. The resisting is performed by establishing individuality through the maintaining of individual space and form and in such maintaining of individual construction it confirms independence in a mode of self protection. The self-protection is a resisting and the resisting the joining of the larger space where such resisting of relinquishing independent space brings about mass. The establishing of a mass factor is part of the resisting losing independence in a self-protection drive from early uniting with the domineering singularity. As the dominated singularity fights to remain independent it is forming resisting and when the resisting is not capitulating such resistance to capitulate its independent form, that is what we then see as mass. That will demand that the mass is brought about by all that the space contains, which is where all the atoms are individually and as a group and as a structure formed by the group where all that are fighting forms a unit and as a unit it is fighting for individual independence

In the commanding space a^3 forming the superior part of the space a^3 that holds lesser space a^3 in the unit as an independent part of the unit, which is included as a part in the same unit that normally occupies space-time as particles or at least by particles with independent space-time The particles hold some space a^3. The space a^3 the particles hold is directly in relation to the particles the containing structure that has to in duplicating in. Every particle in the unit has to fight every other particle in the unit as well as the unit as a whole for space within the time the group puts on the unit. The more space a^3 there is being relevant to the structure in comparison to the structure as a whole and as a unit that forms other relations with other units, which is duplicating in the sense of being a unit once again is relevant to the space a^3 that the structure destroys as a unit. The bigger the space is that the smaller space hold in relation to the size of all the space combined where the combination is part of the complete unit forming individual space-time as a group but still occupying space-time in the group as independent particles while in the unit. In this unit where the smaller individual space remains part of the unit but having to match to the capabilities of other space also in the unit being as part of the unit the less the smaller space will find it able to match the duplication effort compared in ratio to the entire group but still will find it able to focus a duplication standard that holds relevance to individuals in the group. Two sectors emerge in the star where the outer sector advances duplication as the main focus of the star and the inner circle is the sector that place gravity on dismissing space-time. In the inner part the particles focus much more on the dismissing of space-time in the sector as the main focus of the entire unit. The outer part is overshadowed by space-time duplication. Space-time dismissing overshadows the inner part.

The bigger the space is the more the unit as a whole will favour that particular independent to group where the individual will identify its preference in the star by selecting the group that will compliment its achievement. The space will tend to associate with that group that promoted its goals. The larger space will tend to form alliances with those large enough to favour duplication and those smaller but being more solid will sink down towards the dismissing sector in order to select a suitable position it can locate in matching the smaller space needs. The more space a^3 the particle claims as an independent particle in relation to the space a^3 the container hold as a unit formed by the whole lot as one inclusive unit that relates to the space a^3 the container duplicate is relative to the space a^3 the containing structure destroy as a whole unit representing all other space that is combining in the group as representing the group where then the group effort of dismissing space-time in each individual capacity and to the performance what each individual may achieve where such a combined dismissing manifest in a precise centre of the groups space-time. This is the point focussing the dismissing part where the groups dismissing of space-time coming from every atom would gather and present the unit effort, which is in presentation to the duplicating in motion all the atoms form as a group and in all this there must form a balance to gain gravity. As individual occupying space a^3 the atom is an individual container by own dimensions and as such duplicate space a^3 in this regard. The atom resists the dismissing of space- time in which the individual space the atom holds is also included by the confirming the structural form to the atoms relevancy being k^0 in singularity that is bringing on an independent value that fits the particular k in the atom relating to $a^3 = T^2k$, which is part of the whole unit represented by the combined value of $a^3 = T^2k$.

This relevance means that without a specified container producing individual particles that control the specifies duplication and destroying of space in relation to the outer space such a container will apply a diminishing relevancy of space-time displacing that will equal the displacing of a number of 112 protons and no more which are all working as a unit within a confined unit we call outer space and in conjunction what is dismissing in the unoccupied space- time where the unoccupied space-time can withstand. We know that is a theory because the atoms in space can sustain much less than 112. In short the value of the walls serving our three dimensional Universe can sustain no more than that what a possible 112 protons gathered in one atomic cluster may displace.

Inside side this outer container, which we see as outer space there is inner containers being stars that bear the direct relevancy which singularity is applying being inside stars, puts much more strain on the surviving abilities of atoms. In outer space the atom has an own relevancy of seven and the space demand on the atom is only three that the atom must maintain in order to duplicate. But in stars the containing star places a demand of the containing seven plus the space creating three in relation with the time applying inside a star, which are four. Let's put what was said just now in conjunction to the Earth. Since the Earth has no singularity demand that is that much better developed than outer space is insisting on as being a limit in the Universe needs to afford space-time self sustaining of all there is on Earth by a relevancy of **k** to $a^3 = T^2k$ is adequate in that which the atom normally can sustain and leaving lots and lots of space to spare. We call this stat of affairs inside the atom to be quantum physics where we directly associate the concept of quantum with none quantifiable volumes of unaccountable space in disuse. But in bigger units the space-time displacing relating to space duplication presents with much more demand on atomic structures. As the demand of singularity in such units grow some relevancies within the atom come into play and I developed a formula to place such a demand in relation to singularity where the ultimate demand sets the standards. As I stated there is within the star that shift as the star progress through development. At first the star leans heavily towards duplication. Then as the star develops the star moves across a broad range of specifically identifiable stages coming from one extreme where our Gas "planets" (which is stars in the making) is to where a tiny star such as the sun is growing towards sizable monsters and then on to cosmic destroyers.

Through out the variety of development there is a balance unfolding that at first support motion by duplication and leaves dismissing for a senior partner to commit while growing the presence of a superior partner. Then the space-time development allow the smaller star to drift away from the domineering partner as space –time develop amidst all the partners involved including the material giant and not so much giant. The growth confirms the security of individual singularity in the presence of other singularity united under an elected unifying centre singularity. From the motion that points to the stage where atoms dominate as they bring about motion with much less dismissing the stages come and go but the direction of development is always in the direction of centre singularity committing unifying in the extreme by dismissing space-time. In the end the star is totally committed to dismissing space-time by being absolutely unmovable and as it is so solidly stationary it places all motion outside the star into outer space. The aim of this development is to secure all singularity which was at first vested in every atom in independence to a shift towards and eventually including all the atomic singularity into one controlling singularity where the purpose of atoms in independence is taken over by one all including and controlling governing singularity in the centre of the star. All the singularity that was present before is then included in a centre spot that is not even a dot any longer but then it returned to the spot. The black hole is a star as all other stars but the Black hole completed the journey by taking all the atoms in the unit and unified all singularity into one position that replaced all atoms and secured their singularity into one spot.

At first when the star is in a duplicating prone state, the atoms in a range of elements control and produce the drive of the star. But gradually the protons in the atomic cluster fuse into larger units and the larger proton cluster units starts to challenge the drive and eventual destiny of the star by fusing together elements with much less protons in clusters. The working process is immensely more integrate and complex than the process I describe here but this is the shortest introduction I am in a position to give. In short at first in the duplicating stages the atoms take charge of the driving of the star but as progress in development soldier on the centre singularity takes command and the star dismisses space-time while it also fuse together elements in the dismissing process. At first the singularity sustaining the atom has to group together to sustain the star unit but in time the unit becomes string enough to sustain the star without depending on atomic independence to overcome overheating. The main picture arising from the explaining is one of a balance that controls the star through out its development. In the sun for instance which is a minuscule small star a relevancy in the outer region might be all–favouring duplication, which is favouring duplication relating to singularity and with the atom having a sustaining displacement of favouring the electron position there is no actual danger of the atom demising. The star is still all about its duplicating mission. On Earth as in outer space the atomic difference in mass between the proton and the electron is 1836 to one. The electron is displacing the same space-time but compressing the space-time displaced 1836 times less than the proton does.

The figure mentioned is a specific predetermined ratio brought about by the dismissing of space-time in conjunction with duplicating of space-time showing discrepancies between the duplicating factor and the dismissing factor. That means the "outside" is 1836 times away from producing what the "inside" within the atom core does reduce space-time. In the sun this ratio will be much less because the relative mass of the electron will form a relevancy number of 27 where it is on Earth only 3. That is one factor I placed on the ratio whereby the density intensity of the star space-time increases from as the star atmosphere (if you will) of the star such as the sun grows denser and gravity makes the space-time more compact. The density shifted towards the intensity the electron has and as the intensity of the density progress by becoming more compact through gravity the space between particles will match the electron at $(\Pi^2/2)(4(\Pi^2+\Pi^2)/7)=$ **55.66**. This displacement value produces gravity and it produces electricity by intensifying what is in outer space at a value of $7/10(\Pi^6)/6=$ **112.162** to what limit space-time displacement may endure while remaining three- dimensional. The number of 55.6 personifies the maximum proton displacement ability while remaining in form within the sphere. After that the sphere begins to show miss forming. In is no coincidence that that all gravity driven structures has to have a molten iron core and electricity can only be generated by an iron armature in the presence of a magnetic inducting copper chargers producing a field. In spite of the proclamation of science about the sun being only a hydrogen star the concept is nonsense because not hydrogen n nor any other element except iron is able in our Universe to produce through motion the effect of space-time reducing which we either call electricity or gravity. Electricity and gravity is the same ting.

The electron position or in other words the favouring of the duplicating tendency within suns ability on creating a working environment within the sun as that will have has a diminishing factor going as low as 27. This means the atom can reduce the electron space-time occupation by 27 times whereas the atom can sustain as much as 1836 times further than the electron can and the electron matches the speed of light. As the relation in the atom within the sun degenerated by 27, which means it loses to find compensation in the time duration extending by the ratio, which the space reduces in the atom is left with a sustaining value of the electron plus the neutron applying space-time displacement without involving any of the neutron at all. By declining of personal atomic space in order to avoid deforming and losing individual identity which will lead to accelerating early uniting through absolute space-time demising such diminishing that will bring about the relinquishing of individual independence carried by having independent gravity securing the atomic identity which brings over the result that produce mass. That is the mass that the electron will consume in the space reducing and producing an enhancing of mass within the star. When the composition of elements within the star is such that the combined effort of all the atoms reducing space and thereby improving their individual space-time dismissing as they relinquish the same factor in duplicating space-time by motion. By moving the control of gravity from the individual gravity vested in the motion of the individual atoms towards and in control of a centre elected spot holding all invested gravity secured such increase of time by the reducing of space will produce the favouring the dismissing trend much more than a duplicating by motion provided by the compliment of electrons spinning about. When the central governing singularity takes charge of gravity within the star the star goes dark as there is then less electrons in the star and it will reduce the amount of photons flowing a way from the star because what we think of as light coming from a star is the ejecting of excessive electrons not needed by the spinning atoms and such ejecting of an overflow of electron production will have the star shining at night as a bright little boy shining by dismissing pebbles of light-photons into space. It is when the star gets dark that we can know the real monster woke up rise and dismiss in earnest. The star going dark will happen when the centre gravity will request more motion through dismissing of space-time that would light photons have the ability to escape. The proton then takes command from the electron and changes commitment within the star from duplicating in motion to dismissing by excluding all motion from the star.

When a demand of space-time displacement to the value of 56.6 protons becomes the norm the star will seize having space-time concentrated to a liquid by concentration as the star by that time exclude all electron functions and stop shining. This comes into affect when the demand on space-time duplication and reducing the reduced due to mass overload the atom to space without a heat envelope. Only the nucleus will be able to sustain the further diminishing while the reducing of space is directly coupled to the increasing of time. The atom would shrink to such little space it will have space within the star that only the centre nucleus of the atom will take up to fit. More reducing by applying motion in creating space differentiation will leave a star with so little space the space will be insufficient to secure a position for the neutrons and the star will then have the name of being a neutron star. Going even further will find the proton rejected from the star. That is how gravity applies because it is a matter of relevancies applying between space holding and demanding conditions and space reducing in relation to insufficient motion bringing about much less space duplicated. The space duplicated brings about mass as a result. We shall again return to this topic in the new suggested theory later on after much more exchanging of information and arguing about the introduced information has taken place.

In the previous explanation it becomes clear that there is two forms of gravity applying through out the cosmos and not just one form. Saying this I first have to reconfirm what Kepler proved that space could only be when space is in motion and the motion is in relevance to a specific controlling centre. The accepting and the understanding of these principles are absolutely vital to our understanding of the cosmos. This brings across the truth about the expression that one must not think of the heavens in the same terms as we think of the Earth. Laws applying in heaven and laws controlling the Earth by nature are one Universe apart. The Earth serves life while the rest of all the heavens are hostile to life. It even seems to us that the rest of whatever fills the Universe is meant to destroy life. Except that what is on the Earth, the rest of all created has one purpose and that is the destruction of life. If life cannot find any means of supporting life in surviving in a natural state anywhere in the rest of the cosmos out there we have to adapt our thinking about nature in considering the cosmos as totally different from what we find on Earth. We have to accept what we invaded and infested on Earth, but that invasion will be the only part of the cosmos we are likely to invade and infest.

Newton saw what physics applied on the Earth and Kepler saw what physics applied in the cosmos. Kepler saw space is in maintaining space by the motion thereof. The accepting and understanding of this is absolutely vital to our understanding of cosmology. This brings the truth about the way we have to regard cosmology. What is applying on Earth is almost definitely not applying in the cosmos at large. We may never think of the heavens the way we think of the Earth. Heavenly concept stands widely apart form the Earth because the Earth came about to support life whereas the rest of all the heavens do not even know about life existing and is hostile to life. In the cosmos Kepler's gravity overshadows the gravity Newton saw. Kepler saw space is in maintaining space by the motion thereof. In this statement there is a balance maintaining equilibrium of space specifically duplicating by motion in precise equal duplications of the previous space that is repeated by the duplicated space by precisely copying as the following bisect of the previous copy of space to perfect in precision. I once again at this point have to remind that such duplication by bisecting is within the space less surroundings of the proton. It is not in the Universe we see when looking at the night sky. This is what the formula $a^3 = T^2 k$ translates to when turning the written mathematical code to the verbally pronounceable English. There is a balance forming equilibrium on both sides of the divide by producing $T^2 = a^3 / k$ and when barriers are broken and lines are crossed the defining ratio change to $T^{-2} = k / a^3$ where the singularity distance in relation reduces by the time component going negative progressively. What this brings to light is that there is two points forming relevancy which indicate a separation of space although both is sharing in one space with both in the position of the identified space having motion that is balancing gravity by motion.

In the previous explanation it becomes apparent that there is two forms of gravity applying through out the cosmos and not just one form. Saying this I first have to reconfirm what Kepler proved that space could only be when space is in motion and the motion is in relevance to a specific controlling centre. In this statement there is a balance maintaining equilibrium of space specifically duplicating by motion equal to the previous space that repeated a duplicated space by precisely copying as the following double of the previous copy of space to perfect precision. This is what the formula $a^3 = T^2k$ translates to when turning the written mathematical code to the verbally pronounceable English. There is a balance forming equilibrium on both sides of the divide by producing $T^2 = a^3 / k$ and when barriers are broken and lines are crossed the defining ratio change to $T^{-2} = k / a^3$ where the singularity distance in relation reduces by the time component going negative progressively. What this brings to light is that there is two points forming relevancy which indicate a separation of space although both is sharing in one space with both in the position of the identified space having motion that is balancing gravity by motion.

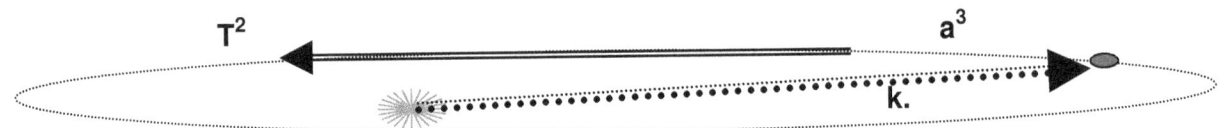

T^2 a^3

k.

a^3 is space occupied by material seeking independence from the centre but that is not all because space a^3 is space holding an identifiable position all the way through the length of the line indicated as **k**. The space a^3 refers to a space a^3 within the space a^3 which all depends on where the motion draws the attention and the space a^3 will only find relevancy when motion sets whatever space a^3 one refer to apart from the other space a^3 referred to. But by identifying one both finds identification because the one is not identifiable if the other is not prominent too. The prominence comes from distinguishing both sharing a joint position.

T^2 is space in motion towards the centre of the space holding the space a^3 that is in motion, which is the space a^3 that is validating the space in motion towards the centre.

k Indicating the distance of the motion of space and in space in relation to a very specific centre. By indicating the point which **k** indicates **k** also indicate the space a^3 becoming the unit of all the space a^3 being in contact with the centre and being in space from within that centre from where **k** indicates space a^3 which through motion is distinctly not the dominating space a^3 that is in motion towards the centre but the space a^3 in motion that is differentiating by distinction separating the space a^3 that is dominating from the centre the space a^3 that is dominated by the space a^3 from the centre and through this motion relevancy the relevancy is holding all space a^3 connected to the centre.

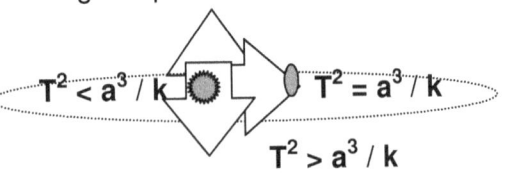

$T^2 < a^3 / k$ $T^2 = a^3 / k$

$T^2 > a^3 / k$

Let us see what gravity is in reality when two objects perform a commune with gravity applying. Gravity applies between space occupying structures in space not occupied and a centre

The space a^3 that is dominating the space a^3 from the centre

The space a^3 that is dominated by the space a^3 from the centre

The smaller space a^3 is distinctly distinguishing the larger space a^3 as the larger space a^3 is housing the smaller space a^3 where the factor **k** is as much indicating by length the larger space a^3 as much as it is indicating the end of the length of the larger space a^3 at the location of the smaller space a^3 by directly pointing at the position the smaller space a^3 holds. Where the larger space a^3 ends the smaller space a^3 is. The two remains as an inseparable single unit in double motion where the motion identifies the unit as much as distinguishing the separateness in the unity and always remain in absolute relevancy.

If space were zero or nothing as Mainstream science so affectively teach us then Kepler's principle formula would need the changes Newton brought about. But it is true and stands tested like no other research ever coming either before or after Brahe and Kepler's work.

From the implementing of $a^3 = T^2 k$ we can see that

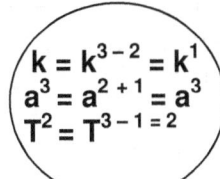

$$k = k^{3-2} = k^1$$
$$a^3 = a^{2+1} = a^3$$
$$T^2 = T^{3-1} = 2$$

$k = a^3 / T^2$	$a^3 = T^2 k$	$T^2 = a^3 / k$
$k = a^{3-2} (T^2)$	$a^3 = T^2 k^1$	$T^2 = a^3 / k^1$
$k = a^{3-2} = k^1$	$a^3 = T^{2+1} (k^1)$	$T^2 = a^{3-1} = T^2$
$k = k^{3-2} = k^1$	$a^3 = a^{2+1} = a^3$	$T^2 = T^{3-1} = 2$
is the same as	**is the same as**	**It is all the same**

$k = k^{3-2} = k^1$ is in direct relation to $a^3 = a^{2+1}$ is in direct relation to $a^3 = T^2 = T^{3-1} = 2$. With this information staring mainstream science in the face and scream pleading at them to recognise this information they turn around and ask why can man not fly off to other galactica at the speed of light
.

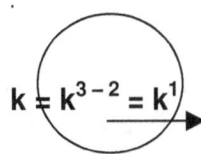

$$k = k^{3-2} = k^1$$

It takes time for space to fill **k** in the distance. In fact it takes the distance that **k** developed since the Big Bang $k = k^{3-2} = k^1$ to fill the distance.
It also takes time $T^2 = T^{3-1} = 2$ to produce the distance forming k^2
It takes space $a^3 = a^{2+1} = a^3$ to form k^3 since coming from the Big Bang

We find that manmade structures in orbit in outer space have a relative very short life and the corrosion up there destroys the material considerably in a relative short period. This is most apparent when comparing such corrosion material decay in Antarctica. In the South Pole articles remain seemingly destruction free for centuries whereas in the desert the heat quite literally dissolves material and even more so is the case in outer space. The heat n the desert as the heat in outer space corrodes material many times more that what is the case in outer space. That means it is not merely **k** but that what forms the concentration forming **k** that also has a strong influence.

When the astronaut is departing from space on Earth or filling Earth space it will take the departing astronaut k^2 time to reach k^1 and fill out k^3. At present and in this moment our most impressive astronautic engineers will devise an engine that would cut k^1 by say half. This achievement will come as they increase the power output say for argument sake to double what it is at present.

There are always two singularities in relevancy. The motion of T^2 seen from the centre in contraction uses the T^2 coming about as the **k** factor for the lesser space a^3 applying motion. Therefore where T^2 is representing motion to the larger **k** it is taking T^2 as the figure that represents **k** as a motion indicator to the smaller a^3. It means that $k = a^3 / T^2$ and T^2 to the smaller a^3 is the **k** factor of the smaller space soldering from point **T** to point **T** which then is the relocation of a^3 by the distance of k. $k = a^3 / T^2$ means a^3 was moved the distance of **k** in the time T^2.

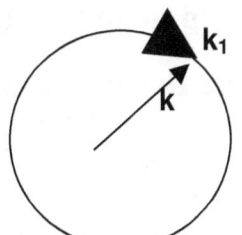

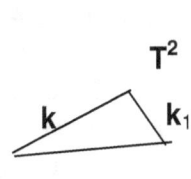

$a^3 = T^2 k$
but $T^2 =$

$T^2 = k_2$

It is conducive to remember that there another part of the two relevancies applying where one is a^3 that is relevant to **k** but also there is the point where **k** has a duty to for fill in relation to a^3

When the astronaut is departing from space on Earth or filling Earth space it will take the departing astronaut k^2 time to reach k^1 and fill out k^3. At the point a^3 then serves the new k will relate as much as it has to adhere to the T^2 time it takes to keep a^3 attending the new orbit T^2. At present and in this moment our most impressive astronautic engineers are devising an engine that wills double T^2. This achievement will come as they increase the power output say for argument sake to double what it is at present. What we see happening in space with objects in motion translates equally to what happens to objects entering the Earth's atmosphere. There is a smaller projecting of space that changes **k** because of an altered **k**. Mainstream science promotes the idea about particles coming into contact with the fuselage of spacecraft when they enter the Earth atmosphere and thus cause friction that entertain heat which then rises as a result of such friction occurring. I know well I had this argument before but I cannot underline the incorrectness more of the way that mainstream science views the principle. By acknowledging this very incorrect way they see what applies when the heat blankets the incoming aircraft will disallow in further accepting of the understanding how and what will apply in such conditions because those condition express cosmology in detail.. There is no friction between particles of the craft and of the atmosphere that is destroying the frame of the craft because it also is true that in outer space there are not enough particles there to bring about such a structural decomposing in outer space or in the atmosphere of the Earth an altitude to do it. What happens is that relevancies reapply and we know science acknowledge that material (they say some but in fact it is all particles) that material reduce the space it occupies when coming from outer space into the Earth atmosphere. In the atmosphere the reducing of space a^3 comes about because T^2 increase upsetting the ratio in $a^3 = T^2k$. When the space in the atmosphere became too small to allow the time it takes to enter because the distance **k** decreased faster than the space a^3 could compromise with the time T^2 changing from what is present in outer space comparing that to the time in to atmospheric space. The space shrink and that push time back into the pas when the heat surrounding objects were much hotter than t is now. Time moves back as space decrease towards the time the Big Bang was present as we know the Hubble concept suggested. With the information in hand for a period of almost a hundred years and where the information forms the basis of modern cosmology since the information formulated gravity and not merely produced a name for gravity as our English friend did it is amazing that such accidents can happen and it is more amazing that no one in Mainstream physics has not the slightest idea or inclination as to why this is taking place! To top the cake with a red-hot cherry we know that our most impressive astronautic engineers are assembling a machine that will scramble the ratio Kepler introduced to a level in outer space where the ratio will be more than what the ratio in the sun is. Surprisingly they are not in the least surprised that not one object in outer space is using an excessive velocity.

An object can rotate in outer space as long as it can maintain a speed that will keep the object rotating in that orbit. The speed requires that the distance from the centre of the Earth to the centre of that object rotating must remain even at all times. That is the gravity applying up there in outer space

When the motion decrease and a lesser motion differentiation sets in the object can rotate in the atmosphere as long as the motion will last and it can maintain a speed that will keep the object rotating in that orbit position. But the difference is that the speed required orbiting declines. That is still the gravity applying up there in outer space

Then the orbiting object slows down to a speed that cannot keep the object in orbit above the Earth and the restriction places the orbiting object on the ground while the orbiters serves it speed as equal to that which the Earth provides. That is once more the same the gravity that applying up there in outer space

That point the object has mass but by increasing the speed the object will increase the rotation speed putting more space in relation to the time it takes to orbit. One may be stupid enough to by into the bluff that the object is still clinging onto the mass it had and that that mass is bringing on the gravity, but as I said that is only when you have a mind weak enough to bay into that propaganda.

Gravity is the maintaining of speed in relation to other motion that is either contributing to the object in motion or the influencing of such a motion.

Travelling is about bringing space in motion. Motion is combating heat and heat brings about expansion. Expansion is producing material as a substance that accumulates by growing into more material and producing material is about duplicating through filling vacant space. To move from one point to another point the material must release from the space it filled and fill the space it is moving into by which measure it then produce material. The lesser the material is, which duplicates in the way of being an individual unit is by taking up space in a larger space unit where the space that is taking up space forms a part of the larger and containing space. The method of filling space as forming a part of the containing space being as such the unit of such a larger space unit. But in as much as filling the larger space the smaller space still hold individuality be motion pronouncing the independence as well as the inter dependence of both individual spaces in the unit. It is the time it takes to duplicate such a large unit when comparing the two individual space units in relation of time each has taken to duplicate, that the ratio between the two duplicating will prove to be of less time duration or "slower" in time duration that the larger motion will take to complete the duplication of the larger space in comparing to the time it takes to duplicate the smaller space. On the other hand the more the space is that is in the process of duplicating the longer period in time such a process will take to perform such duplicating of space per time unit. To duplicate space per measure of space size which is having more **k** that holds time further from singularity by extending the space the "slower" the larger space duplication will be because of the bigger effort it takes to duplicate more space although both uses the same time frame to duplicate. It is not that complicated because a lorry duplicates by motion as does a bicycle duplicate by motion but to get the truck up to eighty km per hour takes a hell of a lot bigger effort than to duplicate a bicycle at eighty km per hour. The motion T^2 requires effort to reposition space a^3 as space a^3 duplicates using time T^2 to shift space a^3 across a distance **k.** The more a^3 there is to extend by the increasing of **k** the more T^2 will be required to complete the task. The faster the duplication is the further the distance is from the centre of singularity and the longer the rotation will be in relevance. To duplicate space using the same time and duplicating a smaller space will bring about a much reduced distance from the centre using a shorter **k** in $k^{-1} = T^2 / a^3$.

We know that science do acknowledge the fact that material (they say some but in fact it is all particles) reduce when the space it occupies is re-entering the atmosphere as it is coming from outer space into the Earth's atmosphere. In the atmosphere the reducing of space a^3 comes about because T^2 increases, resetting the ratio in $a^3 = T^2 k$. When entering the space in the atmosphere the occupying volume of space becomes too small to allow the time it takes to enter because the distance **k** decreased faster than the space a^3 could compromise with the time T^2 changing to bring effective re-aligning. This still is all to do with the Roche principle as space turns to heat to align with the new limit T^2 applying from what is present in outer space comparing that to the time in to atmospheric space. At entering the atmosphere such entering takes the space shrink back to the size and the heat that applied at the time the Earth was released and formed an acceptable barrier with outer space. The space differentiation that started then we now call the atmosphere. By the object entering the atmosphere of the Earth that entering pushes time back into the past. It takes the time in concern, which is the atmosphere of the Earth back to the time when the heat surrounding objects were much hotter than it is now, even in outer space. It refers to a time when outer space had the space-time density we now associate with the atmosphere. By entering the Earth atmosphere it shifts time that then moves back to an era when everything was much more denser than what it is at present. By entering the space-time increases as space decrease shifting the factor **k** towards the position it had when everything was much closer to the time at the time the Big Bang was in full swing and with all characteristics being present then as we know the Hubble concept suggested. With the information in hand for a period of almost four hundred years science completely ignored it all. Where the information should form the basis of modern cosmology it shines in its absence since the time information was formulated even before gravity was named this information was available to everyone that showed the slightest interest. But the interest asked for more clarification than merely producing a name for gravity (which our English friend did). It is amazing that such accidents can happen in science and it is more amazing that no one in Mainstream physics never showed the slightest appetite to research the idea that Kepler introduced or inclination as to why this is taking place! To top the cake with a red-hot cherry we know that our most impressive astronautic engineers are assembling a machine that will scramble the ratio Kepler introduced to a level in outer space where the ratio will be more than what the ratio in the sun is. Surprisingly they are not in the least surprised that not one object found in nature in outer space is using an excessive velocity like moving at the speed of light Oh, I know they have everything confused in the red-shift and the blue shift because again no object can travel even close to the speed of light. The Red and Blue shifts are all about lenses swapping relevancies and that I explain to a certain detail in "**A Cosmic Birth … Dismissing Nothing".**

The concept behind travelling is about bringing space in motion. Motion is combating heat and heat brings about expansion, which is one more concept of motion. Expansion is producing material as a substance that accumulates by growing into more material and producing material is about duplicating through filling vacant space. To move from one point to another point the material must release from the space it filled and fill the space it is moving into by which measure it then produces material by means of duplication. The lesser the material is, which duplicates and by duplicating forms an individual unit, such duplication is increasing the space in duplication by reducing the space through duplication. By taking up space in a larger space unit where the space that is taking up space forms a part of the larger and containing space the relevance of producing more space by duplication which at the same time is reducing the space by halving the duplicating space changes the relevancy the smaller space shows in relation to what it holds in the larger space. The method of filling space is forming space and filling more space as a part of the containing space where the small and the large space is being as such one unit of such a larger space unit. But in as much as filling the larger space the smaller space still hold individuality by separate motion introducing a different $a^3 = T^2k$ in motion and that motion is pronouncing the independence as well as the inter dependence of both individual spaces in the unit. It is the time it takes to duplicate such a large unit when comparing the two individual space units in relation of time each has taken to duplicate, that forms the ratio between the two sharing a unit but not the duplicating in the unit . Even in a unit the individuality one finds will prove to be of less in time duration respectively or "slower" in time duration that the larger motion will take to complete the duplication of the larger space in comparing to the time it takes to duplicate the smaller space. On the other hand the more the space there is that is in the process of duplicating the longer period in time such a process will take to perform and complete such duplicating of space per time unit. As explained the k^2 of the larger space becomes the T^2 of the smaller space and the more k the smaller space produce the more inclining it would have to reduce the T^2 factor of the larger space because of interdependency. The relevancy each one shows becomes a factor indicating the duplication of inter-dependence of the other. In order to duplicate space per measure of space size is to be having more **k** that holds time further from singularity by extending the space of the "slower" or the larger space will use time to duplication. This will be because of the bigger effort it takes to duplicate more space although both uses the same time frame to duplicate as both holds motion in contrast or in different directions but still find equilibrium the unit that has to synchronise the time aspect in order to hold the unit together. It is not that complicated because a lorry duplicates by motion as does a bicycle duplicate by motion but to get the truck up to eighty km per hour takes a hell of a lot bigger effort than to duplicate a bicycle at eighty km per hour. The motion T^2 is seeing to the duplicating by motion but that requires an effort to reposition space a^3 as space a^3 duplicates using time T^2 to shift space a^3 across a distance **k.** The time is the equilibrium factor and the space is larger and smaller respectively therefore the k is different, but also at the same time the k implicates the occupying time factor T^2 and where each a^3 duplicates differently the k it duplicates will be longer or shorter. The more a^3 there is to extend by the increasing of **k** the more the T^2 of the occupier will be required standing in as k to the lesser space to complete the task. The faster the duplication is the further the distance is from the centre of singularity and the longer the rotation will be in relevance. To duplicate space using the same time and duplicating a smaller space will bring about a much reduced distance from point of duplication to the next point of duplication in relation to the centre performing the control using a shorter **k** in $k^{-1} = T^2 / a^3$. Where the two objects sharing relevancy are also using the same time the further development brought about space-time having more space in less time.

Singularity provides space-time but singularity is without space and therefore being without motion that takes up time to complete the duplication of space. Singularity starts at eternity and from eternity all space-time develop. The less the space is the faster the motion will be in duplicating the space because the smaller the space will be in need of duplication. But also the faster the motion is the closer such motion will be in relation to the centre as far as relative duplication goes because the bigger the extending is of the **k** in distance by measure of duplicating it applies and the less space it occupies from duplication to duplication. Gravity is the strongest where space is the least and therefore the time that it takes to fill the space by motion will also be the most in time duration. A Black hole is altogether singularity and a Black hole is all about reducing more space into less space by faster motion dragging time to eventual eternity when space in singularity within the Black hole reaches infinity. The motion is so fast the motion reduces the space into infinity but also drag the time to eternity by the same measure. The time factor slows down so much that the light is unable to duplicate enough space in which time will allow to escape in the space that the light in the space has available. By only having the atom dismissing space-time, as is what happens to most stars in our universe such reduced dismissing will lead to more reduced contraction. That means less relative motion. In the end when the universe will draw the final curtain the final gravity will produce a speed so fast the motion will extend the time duration into eternity as it stretches the time beyond Universal limits and too achieve that it reduces the space, by collapsing all space into infinity. I refer to this action as being in the Black hole but one must remember that the Black hole is the ultimate unifying that all atoms within one certain unit can reach forming a single Unified structure. The atom's final stance is the Black hole that became a massive single atom. In our Universe however having the atomic dimensional qualities, this process of dismissing such space is found only in the atom, which achieves it by applying gravity. Somewhere down the reducing line one find the proton is reducing space into the oblivious by increasing motion to the ultimate, but that is the proton and the star is only all the proton's accumulated efforts.

By only having the atom dismissing space-time, as is what happens to most stars in our universe such reduced dismissing will lead to more reduced contraction. That means less relative motion. The lesser relative motion will contribute to a smaller ratio in the space (not more compact but just less space used to fill) in need of duplicating. With that a shorter duration or period of time will be required to allow the duplicating to come about. The more motion that is required the more in space in the process of duplicating will come about and the further the relative duplicating will be in terms of duplication in ratio to the rest of the surrounding space. This only applies because the relative duration prolongs as the space reduce to comply with the bigger volume of space in need of duplicating.

Speeding up the motion will extend the terms of duplication produced by the motion as the space reduces to extend the time duration. In short: going faster will take longer in time because space reduces by motion duplicating more space per time unit. $a^3 = T^2k$ – this comes down to $T^2 = a^3/k$ and that means extending k which brings about faster motion that will prolong the time duration as much as it reduces the space in motion in relation to the space holding the motion of the space in motion. Every time space halves, each it will take with it the same time and therefore the time doubles through the space that duplicates. The fact that the space duplicates halves the space as much as it doubles the time within the process in duplication. As the space halves each space has an individual alliance with the time therefore as the space reduce it will prolong the time when a quicker or faster motion comes about. Motion is gravity and gravity is strongest where space is least. If an atom is being confined in a smaller reduced space the circle of the atom will have the electron circle growing smaller which will have the electron rotate around the centre core of such an atom. The time is a fixed factor set by the occupying space but with a smaller circle to complete. The same atom will use a lesser confining space allowing the atom more space to be within. The duration the electron has to complete one cycle is the same but when the atom is bigger, The electron travel faster to encircle a bigger circle as it does take to encircle the smaller circle in the same time period. The duration of the spin that the electron will take to complete a cycle will be in the same period as when the circle was smaller therefore the pace the one electron will move about will be much different from the next electron cycle of the other atom in the other lesser confinement. Duplicating space at a faster interval will mean taking space-time back in time which will increase the direction of time to a time where singularity was starting to provide space with time, that is taking space-time back to $k^0 = a^3 / T^2$ $k = 1$ but going in such a direction involves the reducing by measure of $k^{-1} = T^2 / a^3$ to the point where $T^{-2} = k / a^3$. At a point round about singularity the gravity that the space acquire will crush the space the object claims back to the size it would have had, when the Universe was condensed to round about singularity. This cannot happen because long before it happens all space will become heat and the heat will dissolve material into photons. This is the direction we, whom are captured by the Earth on the Earth are heading if not for mass forming to secure our atomic individuality firstly as an atom and then as an atom in a larger unit that is forming a group. Let's carefully look at the general use of gravity as is mostly applied between objects in consent of remaining individually separated by space and with respect honouring each other's independence.

While the smaller (planet) is in a wholesale effort to escape and secure sovereign independence there is the larger partner that is providing the centre from which the smaller object is running. The centre contracts the space it claims and from the centre the object in escaping is as much a part of space-time than the rest of the claimed space-time being the occupied and the holding part of the space unit. Both are relevant as both have a part of space in the unit forming the unit. The centre partner is providing the retraction of motion of the departing object in containing the departing. The second and centre object is retracting the space surrounding the centre object in an effort to supply the object with space the centre reduces through gravity. In relation to the centre the centre is applying motion that is reducing space and the more the space reduces the more heat surrounds the centre point where the space disappear. The space containing the heat disappears but by the space disappearing there is much heat left in the rest of the space as concentrated heat. As the space reduces towards the centre the heat level in that space rises. The centre object is applying motion by dismissing space towards the centre as the centre applies gravity. $k = a^3/T^2$. Then there is the third factor, which is the space itself that is in motion as well as providing motion. This is $T^2 = a^3/k$. As much as the smaller object is running away from the centre, the centre is contracting all the space it claims to be space-time by diminishing the space from the centre. The centre forms a larger space a^3 that provides a flow of spaceT^2 which produces the time aspect that is being concentrated by establishing k being the flow towards the centre k^0 as all the space-time moves the length of k from k inwards to the centre point k^0. From the vantage space holds it finds all space-time equal that it is moving towards the centre of the Universe. The universe I am referring to is the pivotal position as the sun is in the case of the solar system. This we see with light coming towards the observer locating the observer as being in and being the centre of the Universe. I explain this statement in much better detail later on because that statement defines our improper view with which we approach the cosmos. As much as the runaway is running away the centre is contracting the space-time and as far as the centre is concerned there is no special thought going to the runaway because the runaway is all part of the space-time centre but a part which is not that much successfully contracting. The centre is tidying the flow of the runaway but not containing the flow of the runaway. The contracting is successful it is fighting off overheating in a coming together and this that we see we see it to be gravity. The third factor is the space reducing as it is moving and as it is moving and reducing the space by the same margin it is increasing the heat towards the centre by gravity's ability to decrease levels within the decreasing space moving the space towards the centre. $k^{-1} = T^2 / a^3$. We have to accept that rules apply where singularity stands in regard to other singularity. Of the two one is a domineering dominator seeking control as a dominated subordinate fighting off the control by seeking independence. If no working relation is yet formed there is an ongoing fight for position between the two whereby the one will compete to destroy the other lesser developed into submission and the other will put up a relentless fight to flee and secure its independence. I was asked on occasion about my ability proving this statement. Well, we all have eyes and we all have minds so we better use it therefore we all can think about what we can see . We can see there is some dominating going hand in hand with some flight to prevent full submissiveness or a fight to destroy or achieve one of the two relevancies.

In every case it is space-time in motion flowing towards the centre holding a centre spot valid as the space-time is flowing towards the centre and that is providing the motion that is affecting the others sharing the relevancies. The difference (I suppose) between space and space-time is that space is just another meaningless human concept while space-time is having a flow or a motion of a valid substance and such flow is validating the particles or objects and the space-time holding them.   Seen from outer space that motion of a fluid substance is the factor that is bringing about the gravity. Gravity is the relevance of motion of a smaller space putting a movement in relevancy of another moving space within the same space but acting as the larger space while sharing space as one unit with the smaller space and being in the same space. If any reader is in doubt about my statement then tell yourself in all honesty what a force is...but be honest while you explain to yourself what a force is...(the force Newton suggested is keeping the universe glued) and then go and scientifically differentiate in mathematical detail what the difference is between the powers of a Pagan god and a force. From my personal view a force is just motion applying and that is what Kepler said gravity is. Even if you wish to maintain the silly idea about gravity being material pulling each other all over, such pulling demands motion to initiate the pulling or the tendency to apply motion when given a chance to do so.   The pulling starts and ends with motion. The answer about what gravity then is can translate directly from and in relation to the findings Kepler's work produced. In relation to the space surrounding the orbiting structure as well as the space between the centre and the orbiting structure the structure in motion is steady and motionless in concerning the motion of the space, which the centre of the largest sphere is dismissing space towards that centre. The orbiting sphere has a lesser capability to draw space towards such a centre and in that the smaller sphere applies motion in order to secure the maintaining of the lesser singularity in the effort of combating the overheating of the lesser structure.

The smaller object is applying motion by getting through space while in the larger object centre is applying motion to space-time and the space-time is providing the motion linking the relevant object to find equilibrium in motion applied. It is only in this book that I ever refer to space because there can be no such a thing as space in cosmology.

The one object tries to put space in between the centre and the object in a specific time and the centre removes the space between the centre and the object in that same period. That is making the space there is, space-time. That forms a circle and the size of the circle depends on the space relation with the period in time, which produces speed or velocity. There is space through which the occupying material moves. It is at a specific space volume during a specific time period. It is velocity or speed. If the space part is too little comparing to the time part then the time part will contribute more to the ratio and the object will decrease the distance between the centre of the circle holding the gravity applying spot and the object. It is then moving faster than the space is moving towards the centre and the space the object occupy will extend the distance that is between the centre and the object.

What Newton saw, Kepler describes best. $T^2 < k < a^3$ means that the object is falling out of the sky because the time it takes to complete a circle requires much more duplication of space within the space available to the object by the motion for the purpose of duplicating and the space available is not able to provide a large enough period of time to counteract the centres retaining the rotation by restricting such a rotation with an equal contacting.

$T^2 > k > a^3$ says that if the departing object seeking independence shows a greater motion than the distance **k** can provide therefore it will have to increase the space orbit in the time period by establishing the space increase in adding orbital space to establish more space to orbit. By increasing the space a^3 within the new T^2 such extending will force **k** to grow bigger and in so doing provide more space in which to orbit. This is mostly artificial such as one would find in the way rockets are launched but it ring true (although by the tiniest of margins) where the Universe develops by means of extending the k factor. It echoes that which we see in a normal fashion as the Hubble constant and that law describes such expanding. Even comets adhere to time old routes with cycles that are well established and as old as the solar system is. By launching the rocket straight up into the sky following the 7^0 inclination that forms a sphere T^2 becomes **k** and **k** becomes many times the value of a^3 that finally reach outer space. It is a case of this radical increase resulting in more space and with that in the result thereof we find that when the time of the cyclic relation provided by an extended T^2 is to slow and a larger cycle is required because of a velocity ratio that favours the object in rotation, the space between the object and the centre will increase and so will the radius between the centre and the object increase. $T^2 < k < a^3$.

When the space a^3 does not have the ability to produce the required motion T^2 or the increase in speed that is required to accelerate the speed value and the level the Earth centre demands from such an orbiter to remain in that orbit it will seize to provide the opportunity to the orbiter to remain in the orbit. The slowing down of relevancy in speed hampers all further progress of extending **k,** which is enabling the satellite the opportunity the Earth provides to allow such escaping to continue. The shortfall will come from **k** as the length of **k** is reduced and the deducted is in place to compensate for the short falling in the rotating motion that such an inadequate **k** will provide. Then the formula is $k^{-1} = T^2 / a^3$ If the rotating time is smaller than the space the centre provides from the centre to the centre end the distance **k** will reduce and provide a smaller space a^3 in which to rotate T^2 as to establish the required equilibrium needed to secure harmony in gravity $a^3 = T^2k$. This is what we Earthlings experience as gravity, but which is not gravity because it is a bi-product or half the result of the full compliment of gravity. It resulted when some balance went imbalance that crossed the limits of harmony. When the time factor is equal to the space cycle the orbiting structure is rotating within and hold its own in the company of the contracting motion. The lesser space orbiting should claim as much space as the centre is disposing to have equilibrium in space-time. In that the motion providing the escaping has to be the same as the motion providing the equal contracting motion. The motion of leaving is equal to the motion of staying with the circle. It is because of this that comets orbit the way they do and any thought of inter cosmic travel is completely ridiculous. To leave the sun the structure that tries to leave the solar system must beat (not only meet) but beat the gravity coming from within the very infinite of the sun's inner core where the diminishing of space-time provides the space less ness and timelessness needed for fusion between atoms. In order to leave the solar system the craft and all that the craft contains will have to fuse into one atom or dissolve into liquid heat.

By the eliminating of the motion where such elimination is coming from the centre all the space, which the smaller space is within is part of the diminishing and that includes the lesser space that is applying motion. Therefore it is the task of the smaller space to capture space and identify the captures space. When the rotation speed cannot keep up with the dismissing of space and the space dismissing is then overpowering the orbiting space, then the rebalance of gravity steps in where it will try to dismiss the smaller space in total. The orbiting structure will start its descent under such conditions and the orbiting structure will then begin "to fall". If the object is moving more rapidly than the space is depleting towards the Earth the orbiter will "lift off'. But it is all a relevancy of speeds applying placing space in relation to time. It is $a^3 = T^2k$, just as Kepler said. Where the departing speed of the orbiting structure equals the diminishing speed of space in contraction that the centre produce, an orbit of $a^3 = T^2k$ at that point holding time will come about and gravity is in equilibrium. Then gravity in equilibrium departs just as fast as the space holding the departing object diminishes and the departing velocity is the same as the diminishing velocity

When examining the illustration seen at the bottom left the motion in the top illustration indicates as one can see that the motion does encourage the seeking of independence from a centre by a lesser independent singularity. We may take the controlling object as representing the Earth gravity that is securing the object forming the role of the satellite onto a centre that could be the Earth. The reason behind this effort of the lesser-developed independent singularity is seeking independence is because it is threatened by singularity that is the better developed and more controlling. I explain this statement much better a little further on in this letter when I get into the Roche principle. From this securing and the breaking of such gravity securing comes the sound barrier and when such securing border is broken the sound barriers represent a control that becomes invalid. The reason the centre holds the most gravity is by now very well discussed and argued. But the centre is where space disappears and where the Universe goes flat because where the centre is where one will find singularity being the singularity that becomes the governing singularity, which is forming, is forming the centre of the Universe. In relation between the three factors there is the one that is in relation by applying motion within an occupying space. In the space between the two other participating factors there is the occupying space. That space then forms a relation with the space that is roving performing the duty as the surrounding space that serves as the roving occupied space. From any of the centres the whole picture seen from the position end of the line the entire space will be filled by the motion of the space contracting. Observing such space it is apparent that the space in its role as the super container of all proportions will seem to be very motionless. We have to contemplate that the space we regard, as outer space is as big as space can get and the larger the quantity of the volume in motion gets, the more motionless it will seem. By staring at the biggest there can ever be, it will seem to us as being motionless if then only of the shear magnitude that is involved. From the centre nothing is departing but the centre is sustaining the centre by providing dismissing of space that is flowing towards the centre where in the centre the motionless and space less singularity kills off space in motion while the space flowing towards the singularity at the centre is replenishing the centre of space, which is being dismissed by the centre. It is space in motion as Kepler realised $a^3 = T^2$ **k.** Motion is independantly coming about holding three positions independant in equilibrium.

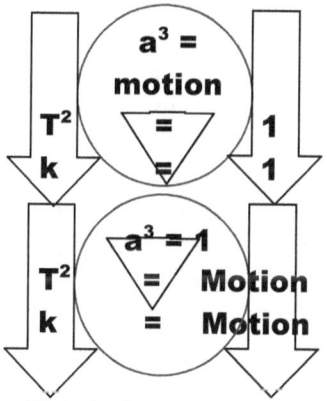

This independence drive to secure independence is part of any structure having the potential to produce and apply heat that will bring about that the second and lesser singularity will search to bring independence to the lesser singularity. It is how the cosmos react to heat coming about as reaction of heat increase.

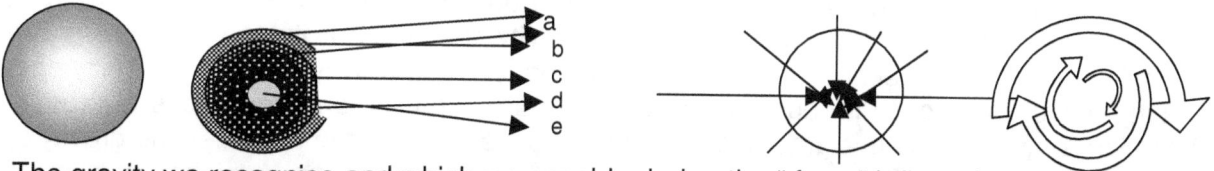

The gravity we recognise and which we consider being the " force" influencing our very existing is the reducing of space by concentrating space to form more dense heat in different layers. The gravity we find in this position is the one sector that Kepler introduced as T^2

The second form of gravity is the phenomenon, which Mainstream Science chose to name momentum but in fact is the second leg of gravity. It still holds gravity as motion but it has a directional change. This factor we normally think of as the one that Kepler introduced as k. In the centre is singularity, which at the time may seem inactive but as soon as conditions re-apply the centre can form singularity that can come alive by motion that introduces the position thereof. By electing a centre point singularity dominating can be anywhere in the cosmos at any given point selected by the motion of all the atoms in a liquid or a solid state or a gas state with the only condition is the asking of relevant motion forming a centre in the application of the

Coanda principle guarding singularity as $k^0 = a^3 / T^2 \, k$ anywhere. This phenomenon we named after Henri Coanda. This statement I shall return too to prove little later on. The singularity stands related to other singularity and all though it forms singularity that group together electing such a centre such a relation can only exist when relevancies to other singularity comes about as space –time or to use another term which is space in motion in relation to the individual singularity grouping to select a centre that will apply to all atomic focusing on a centre gathered by all atoms concerned.

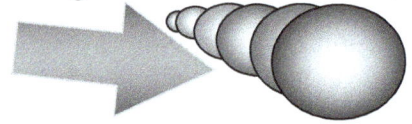

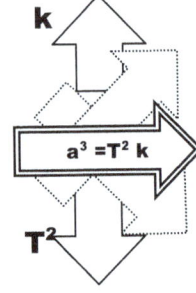

Every relation formed about singularity is about surviving overheating and protecting singularity as much as preventing the loss of individual singularity by destruction. This loss of individual singularity is in the worst scenario forming the Roche limit and in the least scenario securing matter onto the earth. Mass is the product of such a resisting of losing individual singularity in relation to the singularity forming the master or the governing singularity where such an elected governing singularity takes charge of and control all space-time from such a centre and which is from that centre totally dominating all the space-time. This resisting of capitulating of independent individual form giving identifying space by separate motion is forming a position in relation to space occupied by those atoms forming a unit, which we humans chose to the call mass of the material. This does not produce gravity but is the product of gravity. Mass is the product of gravity and not the producer of gravity.

In Newton's gravity formula Newton placed the relation of the two objects in gravity in a square to each other by the moving closer of the particles. But Newton brought in mass as the principle factor the one that is responsible for achieving gravity as a contracting force and according to they're thinking, which is suggested by those most learned in science results in gravity forming. That cannot be the case because in that case there cannot be anything such as micro gravity. Micro gravity is what comes about when mass floats about . It is thought that since the floating object is out of reach of the Earths mass the question is what will then cause the pulling of mass and in what direction will such pulling be heading. So far we have seen that gravity is motion and motion enlarges or reduces mass but mass cannot produce gravity because mass receives gravity. Since we know that which is keeping the objects afloat above the Earth is motion differentiation coming g about as the two crafts harmonise but no equalise their comparing motion or speed. However we find ourselves very much unable to explain the existing of the so-called micro gravity Consider the affect of gravity in micro gravity. It is not micro gravity we meet in outer space but it is micro mass. The pulling of whatever is up there in outer space is less than down where we normally are on Earth. If it is the mass, which is responsible for the pulling of what we think of as gravity, then the gravity can resemble a form of possibly being micro because the mass that is producing the gravity must be the guilty party of going micro. In order to produce micro gravity one must insist on micro mass to produce such micro gravity. Since this argument is childish nonsense we have to realise the gravity up there is coming about from motion where motion produces gravity. A higher motion will increase the space between the centre and the structure while a slower spin will result in the decreasing of the space between the structure and the orbiting structure. The mass remains the same but there is a specific border that an incoming object shall not cross or the crossing will change micro gravity into gravity, which produce micro mass becoming mass. When being up there or down with us the object holds the same composition of material, which we think have as mass in the normal flow of conversation. But the mass becomes a factor not when the specific border is crossed bring the object under the control and in the command of the earth centre singularity. If it did Galileo is wrong and if it does not New ton is wrong. All objects fall at the same rate notwithstanding what mass it has or has not. On route down to the ground there is no mass because all mass fall the same rate. Only when being on the Earth with direct or near direct contact with the earth mass becomes a differentiating deference.

Again I stress that mass as an applying factor cannot contribute to the balance of gravity except for dragging along when being at the bottom and lying on the Earth with the motion of the Earth causing friction between the object and the earth causing what we humans wish to distinguish as mass. Newtonians, you can go on bluff yourself as much as you wish but Galileo insisted that all objects fall at the same rate notwithstanding whatever that mass of the falling objects might be or not be because when falling mass and all differences associated with mass differences are compromised and that discounts mass as a factor in falling. That once again confirms my view that mass is friction caused by the lack of motion contributing to duplication and therefore mass is truly only resistance. According to this explaining mass in contributing to gravity has no role to play as we all can clearly gather from evidence about performance of the objects in the increasing of gravity. Neither has mass any function in this process. It is motion that is creating space. It is about motion $T^2 k$ producing space a^3. With the increase of motion T^2 the factor k would be affected and that will affect factor a^3

$k_1 = k_2$ T^2_1 $k_1 > k_2$ T^2_2 T^2_3

k_1 k^0 k_2 k^0 $k_1 < k_2$ k^0

When gravity is applying a relevancy has to factors equally. There has to be two objects relevancy through the motion they have in positions in space must share space as well as by the independence in the space they share. Such a motion consists of two factors just as Kepler introduced gravity being $T^2 k$. In Kepler's formula there is on the one side space in the cube but at the same time there is a larger measure of space on the other side of the relation. The one side holds the space implicated in the relevancy to a cube but in the case of the bigger space only a line indicate the distance through which the space will measure. The distance is indicating a space that runs from the centre where at that centre there can be no motion as space runs into the contracting end where the motion stops. The space, which is indicated by the running from k^0 to the point cared by a^3 at the end of the line k can only be if such a space is in motion. Kepler said it...Kepler said space couldn't be if it is not being in motion. In that space presented by k the space a3 hold direct relevance to the time T^2 that the motion takes. That means every particle no matter how minute it may be but if the particle complies to the independence it has in protecting independent individual singularity it contributes with having gravity by the motion of $k^0 = a^3 / k T^2$. They have space-time because that is space-time. That is the space located between k^0 and k distinctly pronounced by the applicable a^3 to the tune of T^2. Having those principles has what fills the space qualifying as having space-time.

come about that will affect all and the objects have to form a relation to each other. The two same token share

There at the end is a space a^3 identifying the space that is covering the area running from a specific given centre hat is covered by the length of k. There are within one space a relevancy created by motion of space created by two factors sharing a space. The one shows a negative tendency and performs in an effort in moving away from a centre. Then there is a line that is connecting arch to the centre, which is securing the object by containing the first objects effort to bring about individuality. This produces two points and the T^2 factor accommodates the square relevancy that comes about through motion applying. If the one k comes to soon following the second k the T^2 factor would be too small to accommodate the two a^3 sharing the spot by motional duplication and the object will fall to the centre because of k reducing. When the two points forming k is too large the T^2 will force the object into a larger orbit. If the motion in T^2 is not enough to provide k the required distance k will reduce the space holding the moving of a^3 in place. The circle T^2 in which the space a^3 moves will then reduce in size by placing the distance indicator k into a negative state or a declining measure. That is the gravity we experience but that is a small part of gravity only brought about by motion discrepancy. The object applying true cosmic gravity does not show any tendency toward mass or indicated that mass will affect the falling of the object and with the falling of the object all sizes will fall equally if all sizes and masses in that spot has the same velocity discrepancies effecting all falling objects equally. It proves Galileo correct but it also proves Newton incorrect. We see this with so many satellites and even space stations that plummet towards the Earth. The object did not get more massive and did not through adding mass to either the orbiter or the Earth began its descent as it fall to the earth. Neither did the object become less massive and flew away from the Earth. When the speed of the object goes into imbalance such diverting of the balance occur. The relevancy of the speed balance between the motion of the Earth and the motion of the rotating object changed to accommodate both a^3 in the T^2 that k would allow. That is gravity. That is what Kepler said gravity is when he said $a^3 = T^2 k$. What no one ever took notice of is that gravity acts precisely in the manner Kepler stated. If the motion increase the space increases and if the motion decreases the space decreases as gravity applying is motion in space forming space in motion. The more the motion is the more space is produced and the more space is affected by the increase or decrease of the motion of space. Newton's 4 $\pi^2 a^3 / T^2 = G (m + m_p)$ has no part to play and it is only Kepler's $a^3 / T^2 = k$ that comes into the equation since k is $G (m + m_p)$ in any case.

When k increase all factors has to increase to compromise for the extending of k. If k extends space has to reduce because k is in direct but inverse proportionate relation. In this following argument we find two opposing forms of space where each plays its part in order to maintain a compromise done by both with mutual respect. The one we consider as the lesser trying to escape from the domineering, which is more developed and tries to contain the escapee. With k extending the lesser escaping space a^3 remains just as big as it is but forms a smaller part in the bigger space a^3 being the one in retention of the escaping space a^3. By the bigger space having a bigger area in retention the smaller space is confined to a bigger space while remaining the same space and therefore in relevance by application is reduced in the whole relevancy where it now has a smaller part in the overall enlarged larger space a^3. But in the time aspect the completing of one cycle by the smaller space a^3 within the improved bigger space a^3 that is much bigger and is holding the much longer outer circle of the larger and more of the containing space a^3. The time it takes to circle about a longer space rim will bring about the circling around a bigger space in total using the same time that is taking longer in duration. The roving space can claim more space that will then fall into the space to be concentrated from the centre by the centre as more motion applying to the independent captured roving space will introduce that increase of space into the accumulated space shared by the factors. By introducing more space into the equation it provides a new balance that will suit all the factors in achieving the maintaining balance required. The time component will travel a wider space using the same time component but stretching the duration therefore increasing the time the time used per space unit gained as space holding becomes more but the length of each unit becomes shorter than previously. As the circle increase the time will be adversely affected in duration of space-time. What this implies is that one cannot have space-time and where space increases have time that is not affected by the change in the space.

We gave this forming of separate gravity coming about by means of performing individual motion a name being the Coanda effect to mention one amongst many others. The Coanda effect depends on singularity being a circle and motion establishing an independent singularity. Then the singularity cuts the Kepler formula in two parts. Evidence about this has been with us since the time of the great Leonardo da Vinci whom was the first person to see the potential manipulation of space-time by changing singularity direction by motion.

A low-tech mechanical human device might teach us something about the most basic rules about gravity if we pay attention to the rules as they apply. When a bicycle is motionless and free from support that keeps it erect it will fall down going straight downward towards the Earths centre of gravity. It tips over on a side as it falls onto one or the other side. As soon as independent motion other than that of the Earth comes about in a controlled manner the controlled motion alters the gravity as the motion brings about a balance that establish another form of gravity and is in a way redirecting or channelling the motion from downward spiralling to side ways moving. If the bicycle comes into controlled motion it will redirect the gravity controlling the bicycle. However we should never forget that the bicycle as well as the way the bicycle act The action performed by the bicycle is as artificial to the cosmos as life is artificial to the cosmos. It may be a coincidence that two bicycle builders were the first flyers in the air but it just might not be that big a coincidence. The bicycle represents the first phase required to fly, which is the part just before the part where the object must get off the ground. That is what the bicycle is doping: it is firstly getting the balance of gravity off the ground. The stability gained from motion is much more than what we humans read into it and it has even less to do with human skills

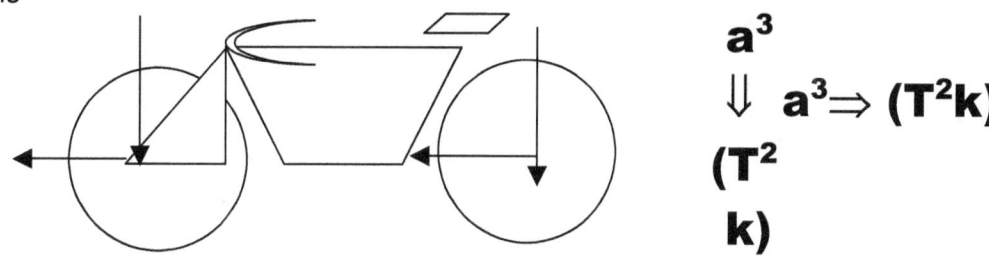

$$a^3$$
$$\Downarrow \quad a^3 \Rightarrow (T^2 k)$$
$$(T^2$$
$$k)$$

A person that acquired the skills of peddling a bicycle has achieved the method of rearranging gravity within singularity. Without motion the bicycle falls on the spot it holds. When the bicycle is put in motion the bicycle can maintain the upright stance as long as the motion applies. When the motion stops the bicycle drops. To introduce motion to the bicycle the motion brings about a stable unsupported upright stance where balance can result from the motion the Earth enforces to the balance coming about by the bicycle using independence gained from motion of the space holding the bicycle because the gravity effecting the redirecting of the Earth gravity response comes about as the result of additional motion that is introduced to the bicycle, This is the very same process that the aircraft need to get air born because it replaces or repositions the singularity the Earth holds to the singularity the bicycle develop in motion. The aircraft only takes the change in direction of what the gravity is insisting on through changing direction in motion through phase one and into phase two. It all is still part of the Coanda effect. With more motion contributing to acceleration the bicycle will become airborne on condition that it is also given the advantage of a set of wings to increase the effect of creating space-time to the advantage of the motion requiring the change in singularity direction.

Standing still with no individual motion   In motion and altering gravity lines with the increase of individual motion

This is where motion through duplicating changes the dimension in equation

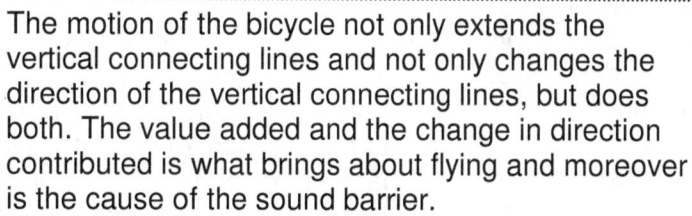

In normally applying gravity we find contracting lines running vertically as the lines connect with the Earth centre. This line forms 7^0 with the centre it of the Earth it connects too

The motion of the bicycle not only extends the vertical connecting lines and not only changes the direction of the vertical connecting lines, but does both. The value added and the change in direction contributed is what brings about flying and moreover is the cause of the sound barrier.

When the bicycle is motionless the bicycle is part of the Earth by gravity applied. As soon as life steps in and bring about separate and artificial motion but that still use the support of the motion that the Earth provide it will inevitably do better than the Earth as long as the motion that life provides is not in conflict with the motion the Earth provide the bicycle becomes an object with the ability to transform the direction of the Earths domineering motion by redirecting gravity there in find the ability in changing the direction of gravity.

The line of the Earths containing gravity is redirected from going vertically downwards to horizontally sideways, which then becomes separate from the vertical line running to the centre of the earth. The bicycle holds a change in gravity flow because of motion interfering with the duplicating of the lines running along the gravity line. The line running according to gravity is in a conflict with the extension that the bicycle motion brings in as the line of the bicycle that implicates the gravity extend in the direction coming about from the introduced motion changing the bicycle gravity line from one vertical running line to a horizontal running line and that changes the line of gravity as it amplifies the line of motion. The line indicating the change of direction of the motion of the space holding the bicycle that is bringing about such motion will then being implicated by the Earths gravity completely redirect the direction of motion into a square in an opposing direction. From going straight down by the bicycles lack in motion when standing still these gravity lines introduce new directions that the line in gravity flowing holds. That then changes from coming straight down to going straight ahead in a horizontal direction forming a new link in relation to the normal straight down vertical and the one committed by independent motion. The relevancies about gravity changed. Every factor receives a new aspect and complete change comes about. By moving space a^3 which the bicycle holds within space a^3 which the Earth holds k changes to take the place that T^2 held as a new k connects k not the sun directly anymore but it now connects to the Earth centre on this occasion while the bicycle is in motion in the manner it previously connected the Earth that is performing in the motion in line with the sun centre.

The Earth takes the position previously held by the sun by establishing the directional motion in accordance with the k^0, which the Earth then provides instead of as previously provided by the sun and while providing k it brings about the space moving When establishing independence that the bicycle holds in space a^3 by providing motion other and above that motion of what the Earth provides in its relation to the sun the object then takes on a directional change in motion. The change in direction also implicate other relevancies as the new motion removes the suns direct contribution as a direct factor to the role of performing as a secondary factor leaving the pivotal contribution which is the major contributing factor then to fall to the Earth in performing from then on until motion of the bicycle stops again as the major singularity centre and gives the pivotal control over to the earth.

The interfacing of the downward motion has to extend as the velocity increase and therefore the connecting point also have to shift promoting the horizontal direction by extending the points in distance of connecting.

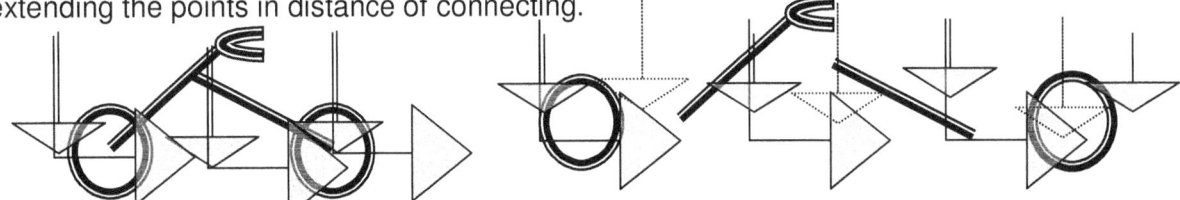

In normally applying gravity we find contracting lines running vertically as the lines connect with the Earth centre. Motion provides extending of the 7^0 establishing the centre connecting points to the Earth it connects too.

It is again Kepler's a^3 = T^2k that changes the gravity relations

When the motion of the bicycle accelerates such points forming connections extend to match to motion. The motion then contributes by increasing the space factor to keep the commitment with gravity valid. The bicycle breaks its form but because it is structurally bonded other aspects concerning gravity has to commit to the breaking of space. When expressed extremely crudely it is put as follows but is very bluntly stated. Yet it still is the best way to explain the basics of the sound barrier. It is the compiling space holding the motion from facet going to facet by duplication together while the bicycle is in the space in motion that is part of the space holding all aspect covered by the atmosphere together in the atmosphere of the Earth. Because the conflict gravity experience gravity first tries to break the object in motion but then extends the breaking to the connecting devices such as the sound waves in the adjoining space The atmosphere does the breaking on behalf of the object in motion since the moving space holding the object in motion as a unit shows much stronger bonding in structure unifying. We experience such breaking of space as the breaking of sound, which is showing motion or gravity differentiation

With the Earth established as the domineering centre controlling the pivotal motion of the bicycle the Earth now presumes in the position the sun had before the bicycle created independent motion and the bicycle now applying motion that produces independence fall into the role which the Earth held on behalf of the earth and the stationary bicycle when all considered the earth and the bicycle was one unit. It still is a unit but holds independence in the unit and even more so than before. Before the motion contributed to independence the bicycle still had independence though that independence came from atomic motion that separated the structure of the bicycle composition from that of the Earth. With the individual motion coming about such motion secures the independence to a next level that will be one step away from advancing to semi self control independence when it starts flying. The bicycle motion will eventually end as T^2 but the relevancy where T^2 now will end is the drastic change that came about. Because the direct control shifter from bringing the sun in as the pivoting factor to the Earth holding the pivoting factor the gravity that is upholding the cosmic equality sense is still upholding the cosmic sense as it still applies since individual space providing motion brings on changes in the factors but not the factors implicating the cosmic law while the relevancies that is required to reposition the structure factors is still in place. The Earth now forms the centre whereas before when the bicycle was motionless this duty fell upon the sun centre. Circling about a fixed and secure centre now bringing about motion that takes the position the Earth has before the bicycle. This is a shift of one position in the gravity relevancies of $a^3 = T^2 k$.

The Earth applies motion in the atmosphere of rotation but from the human position we have on Earth we have to follow the Earth by mimicking motion the motion of the earth. From our stance we are moving in a continuing straight line and the straight line we receive from the Earth is giving us the impression that we are allowing a straight line there is that we can see. Before investigating the principles of flying we first once again must define why the mass is keeping the object secured on the soil. The space the object claims within the space the larger object retains is not enough to duplicate using the time it holds in relation to the volumetric space retained. Not being able to match the retainer the object has to reduce the distance from the centre to secure a more relevant position with a suitable distance to relate too with the duplication of space within the retaining space. This forces the object to relocate to another point where the retainer will match the volumetric space claimed by the retained object and where the containing space is by volumetric size matching the claimed space. As the object moves down to find the correct position of relocating the object is restriction in space

duplication to match that of by a solid structure we call Earth or soil. The solid structure provides a restricting boundary through which the relocating or repositioning object cannot break. The effort of the relocating the object in missing the required frequency has to follow such a search as to find the position of relocation that will match in relation where the specific duplication will match the required density to duplicate the space in the prescribed harmony to the restriction of the retaining object. Because the retaining space is obligated to reproduce a much larger extending **k** from singularity that is required of the smaller space in the same situation, therefore the smaller space is in search of a position all along the line that produces the larger space in an effort to locate a comparing location to match equal duplication resulting in matching factors of space and distance in the same time experienced between the large and small spaces sharing one unit. Such a position will allow the smaller space to have a comparing spot in order to comply with equal duplication when using equal time duration as the time will allow the space then to commandeer a position in the **k** that will serve both sharing the space in the unit. By not being able to find the equal spot in the **k** that will match the space to the duration that the Earth insist on this will force the space to tend to locate while pushing against the Earth and because the soil is solid the soil; will not allow unrestricted entry for the space to go in search of identifying the prefect spot of choice. This relocating and searching for an equivalent position that will match the **k** equal to fit the smaller space in comparing to a point matching the needs of both the **k** factors where the smaller space will fit into the duplication that both objects will have in space in time providing the space that will provide the mass as a factor of control. It is a difference in speed bringing about mass and not mass bringing about the restriction. In accordance with our view we receive because of our position that we hold on Earth our motion is a straight line although the Earth shows a curve but in our minds that curve concerns us little people less. The Earth has motion and by all standards that life apply such motion can never exceed the motion the Earth displays. Our motion serves as an addition to the motion that the Earth provides but we cannot substitute or out perform what the Earth achieve. While flying and duplicating the concept serving mass does not enter any equation at this point. Then later on man found the magic he always were in search of real power in the form of converting heat to space and with this remarkable achievement man found means to break free from the establishment the Earth enforces. Man had more power available to use than just what he could harvest from human and animal muscle power.

This too is the prerequisite to flying by first establishing individual space through establishing separate motion in relation tot hat of the Earth while it is creating motion because the motion establishes antigravity. While the aircraft is motionless on the Earth the Earth motion create gravity and that motion also applies to the craft structure albeit that the aircraft seems motionless to us. The mass of the Earth establishes motion of space spiralling down in time. With the space being in motion the object resist surrendering the form its construction has and with that refuse the accepting of the joining with the Earth by uniting with the solid structure. This resisting action is delivering that what we believe is mass. It is where $k^0 = a^3/(T^2k)$ unconditionally. My argument about gravity being motion becomes prominent when we consider the motion versus space occupied by airplanes using wings to fly. Without motion the aircraft and all that it carries have mass or weight. As soon as the motion overcome the space restriction the Earth enforces on the airplane the airplane loses mass and become airborne. It is only when applying the illogic science use to explain mass where there is no sign of mass present or that mass may contribute to the forming of gravity in any way possible. By trying to correct the incorrect holds mass and weight as a differentiating factor but that does not make sense in any case. You can try and argue till your tongue feels numb but in arguing about the matter no one will ever be truly convinced about that one can believe as being totally convinced about mass still being a factor in outer space. The protons form the centre of the atom and by spinning 1836 times more than the electron the protons dismiss 1836 times the space that the electron does. By having so many more protons in any unit dismissing space a total effort will contribute to displacing more space-time than the effort of a lesser number of protons can achieve. There is a way to detect gravity and that is in looking how we fight gravity or eliminate the effect of gravity securing objects on the ground. We know that $a^3 = (T^2k)$ locks objects in position and flying eliminates our being locked onto the ground. Let's us look at why we fly.

There is very little difference between the bicycle in motion and the flying aircraft except for the motion intensity and by which the duplication of space contact will allow the aircraft to fly and leaving the other object which is the bicycle with much less space contact being without wings unable to fly. The only difference is that the aircraft produces a higher relation with space in possible motion that is contributing to space contact and space duplicating by employing a greater surface to service. In each case the aircraft wings in space contact is providing more space between the two in relevancy where the Earth forms one the object in motion being the space craft or the bicycle. Of the two in motion the aircraft service much more space than does the bicycle. By providing a bigger motion T^2 time factor will extend further providing the space a^3 more opportunity to duplicate then, which the bicycle provides. That will increase the k factoring the case of the aircraft because the more motion will fill more space a^3 and with more space a^3 filling the totality of space a^3 will be consisting of the larger space a^3 to entertain more space-time that is in occupation than that of the smaller space a^3 has space-time to serve this relation as steady as it seems will prove to gain a bigger relation without with out providing more space –time contact by the wings the motion in itself is adding to space in motion without any additional space added in real terms in neither of the two cases only because the motion and the improved wing capacity leading on a larger contact of space a^3 with more of the same space a^3 notwithstanding the fact that the quantity of space a^3 in either case remained the same. This means the other factors in the Kepler formula being in relevancies of T^2 and k will also have to increase to comply with the balance ratio. On the other hand when the motion goes into the extreme and k proves to increase considerably the space a^3 as a factor would have to compromise visibly by reducing the space a^3 it claims. When the contact distances increase the other factors has to compensate to produce the sustained equilibrium. This comes about as a result of the wings that produce more space duplication received through increased space contact with more space because of the velocity increasing and with more motion as well as a bigger contact area more space becomes involved, which is promoted by the speed factor by discriminating in favour of the flying device to get the aircraft flying. The space ratio by duplication increases in ratio to the dismissing ratio by the combining effort of all protons within all atoms that is forming the flying machine and the machines cargo. However the fact that motion is in place as a cosmic event is totally artificial by cosmic standards of motion. In both examples the motion is artificially applied as a result of life's' extending life influencing. We must realise that the motion we find in the action of the flying machine is not cosmic driven although it is an interpretation of a normal cosmic occurrence that will take place but under much different circumstances than the manner life enforces the motion at the time the action that is brought about as a result of life's obtaining the manipulating abilities to translate to human achievement and as far as life is finding a means to manipulate a cosmic law to increase the benefit of life accept that the action is totally artificial by cosmic standards. There can be no such natural motion where a rock starts flying because it has received from some U.F.O. out side source a set of perfect fitting wings. Judge responsively what belongs to the cosmos without life and what life can reproduce in spite of the unnatural state of such duplication might be. It is most important to realise before classifying and grouping this normal physics action inspired by the intervention of life that there can be no such motion coming about on a planet without the presence of life bringing about artificial motion. One can inspect the moon all you like with the best telescope available to man but one will not see any flying object zig zaging the moon surface. No bicycle can by own initiative come into an upright position and start moving on two wheels. The motion is not cosmic inspired and only by seeing the difference can one have the mindset to venture into the activities applying within stars. Let's see what is artificial about life in relation to how the cosmos relate to life. Mainstream science hold the opinion that life in the cosmos comes at a dozen a penny with change repaid. What a lot of crap his idea is and I mean dirty crap. Life is alien to the cosmos while the cosmos is fiercely hostile to life being an alien in the Universe. Even on our planet life has to obey certain and very specific conditions in some cases otherwise the Earth as friendly and nursing as it is will bring about life's demise. One should try and live a thousand meters below sea level or ten thousand meters above sea level and watch your personal demise comes to you.

Let's venture slightly away from the cosmos by trying to define the role of life as the only force found in the cosmos. Life according to my personal defining is absolute managed heat within specifically designed cosmic fibre having the ability to apply forces of a wide variety giving life power or a valid force to manipulate space-time by manipulating some motion or rules applying on motion thereof. When the body holding life Is not hot or with very low intensity heat it is not with life. If the body has no motion of any sorts it is not with life. If life lost the ability to manipulate gravity in the form of low electricity life has lost living. Life can create motion by manipulating space-time it occupies or which it can control or manage. The only place in the Universe known to man, who is not absolutely in all respects completely hostile to life, is this blue dot we waste for gaining money and profits. We should never confuse life's ability to accomplish with that we associate with cosmic events because an apple falling from a tree is life's manipulating motion because it needed the intervention of life to get into a position to fall from the position it took being in the tree and that is not a cosmic event. If the apple came from the outer space it would have been fried charcoal before it reached the Earth and that result is a cosmic event. That part is the part everyone in science including Newton and Newtonian disciples ignores or chooses to ignore.

We by which my referring includes most forms of life that has the ability to stand independent on Earth and from the Earth stand on Earth above the very top layer of soil holding our space in the space of the Earth. We cannot have independent excluded space if we do not fill the space we have on Earth. While being on Earth my position is $a^3 = T^2 k$ where k is because of the mass in movement standing in for k^0 by being k^{-1} Since my body duplicating is less than that which the Earth has to duplicate but is confined to the time on Earth all the same such a body will forcedly find that the distance from the Earth controlling singularity such a distance is corrected constantly as to fit into and apply with the standards that the Earth standards insist on. However with me calling the Earth space-time a force whereas it is the normal gravity flow the force comes from counteracting the flow of gravity. Being k^{-1} we are also T^2 / a^3 which is reducing us in the space we hold and that is only our mass that comes into affect as the Earth repeatedly insist that we try to reduce a^3 further to comply with the T^2 the Earth is applying and which we have to use without any further options given by the Earth. If we wish to confirm our independence of the space we have within the space of the Earth which contains the space we hold by moving through the larger then we have produce a larger k factor by extending the normal k the normal k factor we receive from the Earth to the order of at least k^1 to find the ability to move from k^1_1 to k^1_2 which will allow us to enforce our own gravitational force in spite of the Earth's much stronger natural flow of space-time because we use T^2 to move from k^1_1 to k^1_2. So we have to improve both our independent position T^2 as well as k to accomplish motion. But that puts Kepler's formula in question. Using $a^3 = T^2 k$ and producing a larger $T^2 k$ it means a^3 must also improve. That it does by doubling the space it use during the motion. The space a^3 becomes the next space a^3 because the motion $T^2 k$ is providing the way that will bring about the matching duplication the motion contributes. This is not that uncommon physics. A car holds the space a^3 and is moving by T^2 through the distance of k When the car speeds up to a higher velocity the gravity will increase on the part of the car because the distance k will increase. With the mass or space in motion that remains is not able to remain even with no increase to the actual material used to move nevertheless the potential mass is increasing by the square of time where that is the gravity or the time by the square that increases. The increase in space is the producing of more of the same space by duplicating the same space more in the same virtual time. It is the motion providing the material a duplicate value of its mass (duplicating because of the square used by time) that then forms the increase in the material mass, which our human instincts of sensationalising prefer to call the momentum of the object. But that motion is that what gravity is. The way gravity is applying is acting in the same manner everywhere but man has subdivided the concept under so many names given to misrepresent each fragment of the entire concept unity we divided that we cannot even find the basic principle any more. Gravity is not a force as Newton suggested but a motion that is formed by a natural flow of space-time between space occupied and space waiting to be filled and when filling it's forming a relevancy and this applies throughout the Universe. The only force there is can only be found on Earth in the form of life. Life is the only force and only found on Earth. In spite of all absolute madness that most of the important persons in science whish to propagate in they're apparently attempting to promote an even more mindless concept, which is atheism. They try so hard to pretend that life is a natural flow of normal cosmos that they go as far as to show how mindless ideas they truly come to conclude. Without a God it means that life which is a God linking factor, must be in abundance and if life is that plenty everywhere it is as common as star dust and then it has to be so commonly found we will trip all over other life through out the cosmos. Until proven otherwise we find life on Earth and nowhere else and that fact are written in rock in spite of all idiotic atheistic gibber. In life we find a force different to the letter in the minutes detail to any other factor there is in the cosmos at large. Only on Earth there is life being a force with the ability to manipulate space-time, which is placed under life's control under its control by providing motion other than and above the motion the cosmos does provide in order to maintain and sustain space-time. We use the Universe as if the Universe was meant for us to use. We increase what the cosmos gave us to use as if the cosmos was created deliberately for our purpose and for us to use which is just as corrupting madness as is the jabbering nonsense promoted as the religion called atheism. It is precisely in such a manner that light use to accomplish moving ability to travel in from singularity to singularity. Because singularity in space is space in darkness we consider the space we see as night as dark and therefore invisible. The darkness is light outperforming visible light by duplicating much faster than can visible light. Darkness is light which breaks down and rejuvenate space much faster than light frequency can the photons find a way to escape from the gravity applied by specific singularity points. Being in another frequency of duplication can the photon manage to secure its escapes. With the escaping that the photon does the photon can release and join the next singularity in the period being in a position where singularity takes charge and survive by galvanising a small portion of the heat forming the photon and by singularity releasing some part of the overall heat by removing space-time and forming the motion from the previous to the next infinite position in singularity in rejuvenating the point which is representing by producing space-time, which then will include the photon reassembling with the next singularity forming the space-time of the next singularity. Looking at the issue in this way we can begin to appreciate that light is the duplication of the photon by the singularity charged by the motion that provides the singularity by charging the intensity. The flow of light is about duplicating more than dismissing although dismissing does form part when the photon changes singularity. In that way the light loses intensity to the singularity that releases the light when the singularity releases the light. This process reduces the intensity of travelling light as it travels and is recharged by the singularity on route to somewhere in the future.

In contrast to the duplicating of light is the duplicating of material is more intense and more profound. The duplication of space filled with material is the use of heat compacted in space selected from the surrounding space in the atom forming a unit, which provide the material the ability to confirm the space they hold onto the space they move into without conforming or giving up ground that is filled atomic space, which is much more than just singularity. It is singularity that is sustaining more heat than the singularity will ever require and much more than what particle ever will require. Singularity empty of material which can take charge of light can generate motion to duplicate the photon whereas material use the heat the photon provide when the proton is clashing with the atom. The heat provided by the photon is only a part of the total heat that is required by the atom to replace the dismissed space-time that the atom needs for duplication as well as dismissing of space-time at the centre. The singularity placed in charge and inspired to dismiss has the task to make space-time redundant whereas the electron is in charge of the factor that is making the duplication of space-time, supported by space-time protected by motion in contact with unfilled space-time then even requires more than what the photon can deliver because that is why there is shadows forming the dark side. Looking at this in a clear and sober perspective we once again find a reason to believe that heat and light is the antimatter that matter ate all up and still wanted more. Material is still eating away at light as it did when space began and material still craves for more light, which is heat. We once again find a reason to believe that heat and light is the antimatter that matter ate up and wanted more. Material is still eating up all the light and then still wants more.

Man could create motion but at first such motion was far less than that motion which the Earth provides. The motion of man's ability was vested in what his muscle power could provide. But a very short while ago man grew wise to machines and the fact that machines can provide more motion much faster than could animal muscle bring about motion. By supplying machine motion it gave man extra ability whereby man extended the relation between what the object has when in normal contact with space and when extended by extra motion allowing more space to apply to the surface of the object, thus enlarging the object surface in the relevancy brought on by motion.. Then man found means to break the barrier that muscle strain held and was able to apply motion equal to that of the Earth spinning. After that eventful day then came the day man had more motion than what the Earth could provide. This is where nature and man parted their straight line common sharing of the factor **k** because with the motion that man could produce placed man in a position where man was able for the first time to outperform the earth ability of duplication by motion and thereby can go in disrespect of Earths straight line motion. This is where locomotion generated by steam concentrated gave man more than what man could tap from life and sweat. From then on man produced his very own straight line gravity in such abundance that mans gravity generating ability is no longer in harmony with the Earth's gravity ability and with mans very own straight line man could eventually leave Earth altogether. But let us get back to the straight-line man moved in as that straight line no longer followed the Earth spin. At first with mans first attempts it showed a diverting form the straight line of the Earth and we even gave that diverting a nice name as we do with all things. We called this diverting flying and flying is proving what Kepler said what gravity is in so many ways again and again. The space grows as we increase the motion with which we travel while complimenting the space through which the space travel in which space we are in locomotion.

If mass was the major factor in generating gravity, the mass will play a major part in the time it take a body to fall from any given point to the surface of the Earth. It just has to because the mass then does the pulling. Since Galileo proved otherwise the concept of mass being the producing factor in gravity comes across as rather less thought through and more than a bit silly. In flying a certain criteria must be met by involving the motion of space and as that motion is running through space in time the motion is relating to space by the scale of time. This then has to indicate that there is a restriction in all motion running through space. That we can observe by the speed of light slowing down in denser space than what the speed of light is in less dense space. When an object travels through space the density affects the motion. The slowing down by relevancy as a factor comes about by bringing into the equation where a smaller space must negotiate travelling through a denser larger space. By being in a "thicker" density the motion has to manoeuvre through more restriction say in comparing the space-time we find as the Earth atmosphere being denser than the space-time that the Universe grant. In denser space-time there is more reference points being spots of singularity to contemplate as space distances where the singularity concertina in the time given to fill the more or the less space being contemplating. By lesser dense space there is lesser virtual singularity forming less restricting to the moving object. It is very obvious when looking at an object that is coming from outer space into the atmosphere of the Earth.

As the factor one has to consider that the space represented by the line k increase in the density it represents there is more space to relate to motion. It will be the same as if the space a^3 is moving faster because of relevancies re-applying to conditions changing. Then space a^3 would reduce in the relative factor presence within the larger and newly introduced k that comes about. Then that means that the faster a^3 travels the more a^3 will extend k away from k^0 and by that increase in motion that increase will also apply to the time relevance by reducing the space a^3.there will be a smaller a^3 because there will be more space that k has coming from the being in the larger k.

A larger k will bring a relevance that reduce the space because the holding space increases in relation to the lesser space occupying a part of the holding space. There is always a double relation to space, which is the space the travelling object occupies and the space in the circle of time that is in control of the object by motion because the object is in motion. But this relating is affecting on both sides of space and because of that mans quest to travel at the speed of light is totally unrealistic. Those comedians hiding behind science from where they are trying to pretend to be all the wise about the Universe because of they're accomplished scientists has a smaller chance of going even one tenth of the speed of light than does Little Red Riding hood has herself a bigger chance in finding her talking wolf with the ability of eating Grandma wholesale as one unit than they're having any chance flying through space and achieving 3×10^5 km / sec. In the fairy tale Little Red Riding Hood has a bigger chance in finding her talking wolf than the chance any cosmic traveller will have to go into space flight and achieve 3×10^5 km/sec.

By establishing motion and such motion is bringing about certain contact with space by increasing motion such an increase can bring about much more space to be in contact with the moving structure and rebalance the dynamics of such space in motion. The space a^3 in motion T^2 will establish a larger area k, which is then contradicting the gravity of the Earth's motion in descent and this will count for a larger area present in a smaller time relevancy. While all this is happening the result is that a stronger motion line in a 90^0 directional change will come about and since the Earths motion remains the same it gives the flying structure an advantage to increase the relevancy in favour of the accumulating the dynamics of the balance in space-time by space contact increasing by the contribution of more motion above and on top of that of the Earth motion.

Gravitational Constant
(Symbol G) It is the constant that appears in Newton's law of gravitation. It is the attraction between two bodies of unit mass at unit distance apart. Its value is 6.672×10^{-11} N m^2/kg^2

when the distance is expressed in metres and the masses are in kilograms. Although it is described as a constant, in some modes of the Universe G decreases with time as the Universe expands (see Brans-Dicke theory), but there is no evidence for this.

Gravitational Field

The region of space around a body in which that body's gravitational force can be felt. Within this region, other bodies will experience a force of attraction that diminishes with distance from the body.

Internal Mass

Inertial mass is a measure of a body's resistance to change in its velocity or state of rest. Inertia is a direct property of the mass of a body: The greater the mass, the greater the inertia. Although mass is formally defined in terms of its inertia, it is usually measured by gravitation.

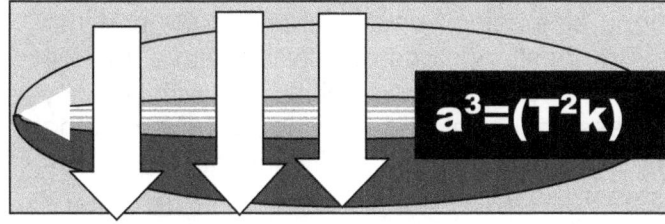

Mass is the refusing of any object to dismiss the form it has and to join the Earth solid structure. Mass cannot and do not contribute to the establishing of gravity except by depleting space through motion of and such numbers of the protons in a space forming an exclusive unit,

Kepler answered the question of flying and airflow dynamics before Newton gave us a name for gravity. Kepler said the motion of space must form equality to the motion of the space. When the aircraft is maintaining flight height the motion equals the mass and equilibrium ensures a constant flight height. As long as the speed and the mass are equal the aircraft will be in equal gravity balance.

Gravitational Mass.

To overcome this breaking effect that the smaller object has on the larger object that we named mass of bodies being on the Earth the smaller object firstly has to transform and transmit singularity from the centre of the Earth to the Centre of the motion wherefrom a new balance sets in. This we call the Coanda effect and it works either on the linear aspect of gravity or it works on the circular

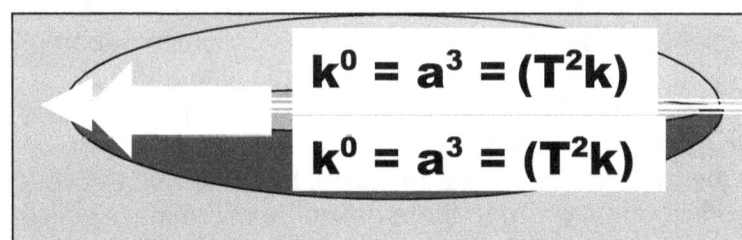

It is when the motion exceeds the mass the aircraft has the ability to brake the sound barrier. Galileo proved that no mass is present in falling, which is also matter in the process of flight and because of that can the sound barrier become some form of constant.

When the aircraft increases its motion, the motion changes to accommodate both space a^3 in the motion thereof T^2 that the applying k factor would allow. The space a^3 is a fact influenced by the Earth but directly dictated by the atoms forming the aircraft. The time factor T^2 is directly derived from the motion the earth dictates. The Earth rules on the distance that k would produce. That is gravity. That is what Kepler said gravity is when he said $a^3 = T^2 k$. That is what no one in four hundred years cared to take notice of or refused to recognise. What no one ever apparently saw what is that gravity and recognise that gravity acts that is precisely in the manner Kepler stated gravity is. If the motion of the aircraft a^3 increase the space increases, which the aircraft influences but also the space influencing the aircraft changes bringing in new alterations. If the motion decreases the aircraft space relevancy increases and if the space decreases the motion changes the contact in space therefore it changes the volume of space which is reducing the relevancy of the space in motion which is gravity applying by motion in space forming space in motion. The more the motion is the more space contract is produced and the more space is affected by the increase or decrease of the motion of space. The more space that is duplicated the more space is produced but also the more space is reduced through such duplication. Seeing gravity acting in this manner does make nonsense of Newton's $4\pi^2 a^3 / T^2 = G (m + m_p)$ changing of Kepler's formula as it has no part to play in correcting the formula of Kepler and it is only Kepler's $a^3 / T^2 = k$ that comes into the equation since k is G $(m + m_p)$ in any case because $4\pi^2$ indicates an individual structure encircling the sun centre when one use the cosmic relevancies which I later introduce.

We gave this forming of separate gravity coming about by means of performing individual motion which is disguised as a very well known and commonly occurring phenomena which is burdened by carrying yet another name of being called the Coanda affect. The Coanda effect depends on singularity forming when a solid and a liquid is in relevant motion where such motion of either the liquid or the solid or both factors has to move and such moving contributes in selecting a singularly centre point that will secure the control of the space-time, affected by such motion. The Coanda principle forms a circle and motion establishing an independent singularity in such a circle centre. Then the singularity cuts the Kepler formula in two parts where space is following motion and motion leads space. Being as human as the next person and showing as much human tendency as anyone else being human I changed the partly to the Coanda effect because that is what humans do best. Humans decide they have no idea what they discover and hide what they don't know they discovered behind But with a fancy name other meaning behind the discovery gets less important and the naming becomes the accomplishment a name will scare away any one also in mind of finding out what was discovered. Like calling heat plasma when plasma is the same as heat gone liquid. Well in my case I use the name as the Coanda effect because it is a process where motion is having an effecting on space-time.

The prerequisite to flying is creating motion because the motion establishes antigravity. The motion reduces the friction that mass creates and in that reduces the gravity that creates the mass. While the aircraft is motionless on the Earth the Earth motion create gravity and that motion also applies to the craft structure albeit that the aircraft seems motionless to us. The mass of the Earth establishes motion of space spiralling down in time. With the space of the earth being in motion the object resist surrendering the form its construction has and with that refuse the accepting of the joining of the Earth solid structure. This resistance we believe to be mass. It is where $k^0 = a^3 /(T^2 k)$ unconditionally. My argument about gravity being motion becomes prominent when we consider the motion versus space occupied by airplanes using wings to fly. Without motion the aircraft and all that it carries have mass or weight. As soon as the motion overcomes the space restriction by defying gravity affecting the aircraft to a stand still which the Earth enforces on the airplane the airplane loses mass and becomes airborne. Notwithstanding the corrupt argument Newtonians bring in about mass remaining a factor. To prove their corruption in this argument, let those that disagree with my stating them being corrupt answer the following. On their admitting we know that mass increase as gravity in stars increase. The more the gravity is the more the particular mass will be. But then the very opposite is true where in space there is micro gravity. Then there has to be micro mass which means by their own admission, mass disappears. It is only when applying the illogic use of mass and weight differentiating and insist on proving the incorrect correct in using a method which in any case that does not make sense by any standard of arguing that one can argue about in order to prove the nonsense about mass still being a factor. Thinking about this I feel delighted that those being so very incoherent about mass see my argument about space and nothing being incoherent. The protons spinning are supposedly bringing about the mass. The protons form the centre of the atom and by spinning 1836 times more than the electron the protons dismiss 1836 times the space that the electron does. Because there is an increase in contact with space by the body/s in motion the dismissing of space-time does not only become fully substituted by the duplication but also totally overwhelmed by the motion. It is the dismissing effort applied by the combined unit of all the atoms in the motion in relation to the contact made that tips the balance. There is a way to detect gravity and that is in looking how we fight gravity or eliminate the effect of gravity that is securing objects onto the ground. We know that $a^3 =(T^2 k)$ locks objects in position on the ground and flying eliminates the flying device including its cargo being locked onto the ground. Let's us look at why we fly.

The Earth applies motion in the atmosphere of rotation but from the human position we hold on Earth we have to follow the Earth in motion. From our stance we are moving in a continuing straight line and the straight line we receive from the Earth giving the impression of a straight line for us to see. Before investigating the principles of flying we first once again must define why the mass is keeping the object secured on the soil. The space the object claims within the space the larger object retains. This then is what should be overcome to fly.

When the aircraft is gaining lift the motion exceeds the mass and with that is adding heat at the bottom of the wing to create more mass a^3 added than the speed (T^2k) can create motion above. Below the wing there is more space in contact with material that is improvising to dismiss more space by collecting space compressed with heat by restricting more of the motion. On the top of the wing the motion that accelerate the flow of heat and by doing so dismisses the possibility of having more space-time dismissed as is the case applying at the bottom of the wing. In that way the wing is at the top creating an environment, which favours extensive duplicating. At the bottom of the wing we have $a^3 > T^2k$ and at the top of the wing we have motion outranking space accumulation by restricting motion therefore changing the balance on that side to $T^2 > a^3 / k$. As the speed gains, the wing will strike a balance and at a certain flight height the motion will equal the dismissing going on and equilibrium ensures a constant flight. The motion of the craft establishes individual gravity that is surrounded by the Earth gravity but the independent motion grants the aircraft some individuality and exclusivity.

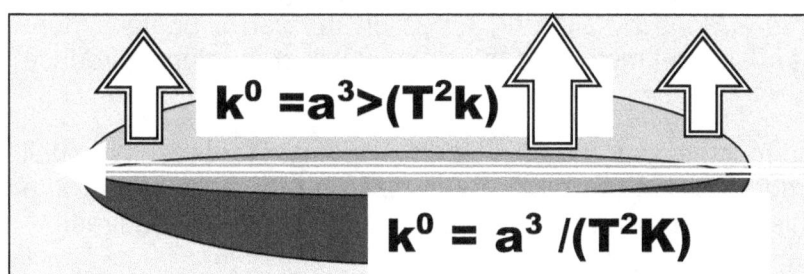

$$k^0 = a^3 > (T^2k)$$

$$k^0 = a^3 / (T^2K)$$

The establishing of independent motion of the craft secures an individual gravity and such individuality leads to the breaking of the sound barrier because the one gravity can no longer subdue the smaller motion, which is producing gravity

By decreasing motion the mass of the aircraft will tilt the balance towards favouring the gravity the Earth applies and the favouring of the dismissing factor of space-time, which then overcomes the duplicating effort of space-time by motion will contribute to the descending. In order to apply a perfect controlled landing the wing must establish additional space-time dismissing to allow the steady descent and the perfect landing. Even when performing the landing under the most stringent conditions the balance still rely completely on the balance Kepler gave us of $k^0 = a^3/(T^2k)$

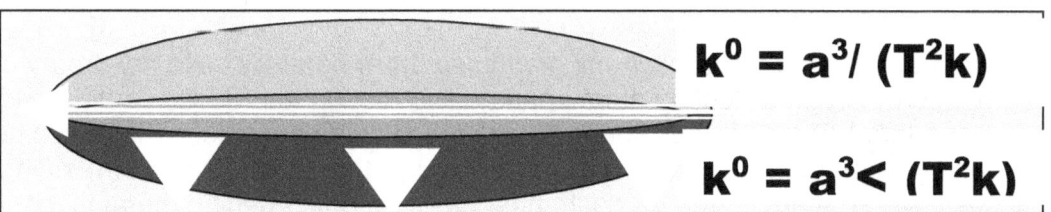

$$k^0 = a^3 / (T^2k)$$

$$k^0 = a^3 < (T^2k)$$

At a height of 31000 km above the Earth the mass of the wing becomes compensated only by a motion of a relevancy that comes about at 2500 km per hour. In that case the craft has to apply motion at a rate of 2500 km / hour just to create the required velocity to keep the aircraft in motion in the sky. Motion creates gravity just as Kepler said when he said gravity is about $a^3 = T^2 k$, which translates to the dismissing of space and the motion, duplication establishes a centre that controls the balance that the newly secured singularity will provide. When the aircraft stands still the sun provide such a pivoting centre but when independent motion comes about the point shift from the sun to the Earth centre where there is a line contact between the singularity that the Earth holds which then forms a new relation in respect to the singularity activated by the independent motion of the moving body which the aircraft takes on motion that the relevant singularity is claiming are released to the minor space. The Earth provides a point from where space depletes completely within the centre of a sphere from where gravity is securing the centre spot in the form and the space surrounding the form that controls the space and time in which the independent object moves (in this case it is the aircraft). When a balance comes about between the departing object and the space reducing only then does an orbit establish a balance of speed serving time duration and space dismissing evenly. That is gravity and that produces gravity only when motion creates a centre to form a sphere.

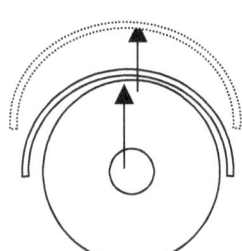 If the **k** that is applying to the distance the space a^3 being in motion T^2 requires _in order too_ satisfy singularity k^0 _the flying of the aircraft is then_ unequal _to the motion in the previous relation that was in place between the sun and the Earth and_ the motion will bring a correcting in the distance **k** to put the motion in balance with space. Space will always demand a correct establishing of the miss interpretation of the equilibrium that is needed to sustain the effected singularity because of the space-time factor.

This is very important if one wishes to understand the sound barrier. The two positions in **k** depends on the motion of T^2 in relation to the **k** the object is maintaining while the object stands in the space a^3 and related to the motion the Earth provides.

When the motion amplifies the status of the object elevates as the motion increases the relation to singularity, which is located in the centre of the Earth k^0. The more motion that applies the higher the velocity of the flying object will be and the bigger distance **k** it can sustain. If the sustaining of **k** is not required, as is the case with racing cars the space a^3 needs relative improvement and that is done by allowing wings on the car to contact more space improving the dismissing part of space-time where this contact will proportionally reduce **k** by amplifying the relevancy of T^2.

In our normal posture we are moving much slower than the Earth is spinning for reasons I shall come to later on. With our bodies not applying the motion we apply the thrust to reduce the motion the Earth has in order to achieve the much-needed equilibrium the cosmos requires.

To overcome this breaking or restricting effect that the smaller object has on the motion which the larger object provides and that we named as being the mass of bodies being on the Earth as being in the role of the smaller object firstly has to transform and transmit singularity from the centre of the Earth to the Centre of the motion wherefrom a new balance sets in. This we call the Coanda effect and it works either on the linear aspect of gravity or it works on the circular aspect of gravity. By applying motion either to a liquid in the presence of a solid or supply motion to a solid in the presence of a liquid a new point in singularity establish a new centre elected by the motion creating the defining of the space a^3 that is in ratio equal to the motion thereof T^2k bringing dominance and the process applying the rules I shall explain a little later on.

All the principles that I make use of to explain my theory is part of nature. I base my theory on heat becoming stabilized through collecting more space using motion to produce cooling. This idea is most basic and that I admit. It may sound basic, but Mainstream science is also most guilty of their departing from this most basic principles through the employing of terminology and terminology has covered many of the crudest, most basic meaning behind the most basic principles in nature. I do not applaud a principle Mainstream science underwrites in the sense that matter in the beginning was coming about and anti matter came to destroy the matter. It is moreover the disappearing from the Universe of the dissolved by product which antimatter somewhere not in the Universe. It is this vanishing being the result of between the two opposing materials that I strongly reject. If anything ever was part of the cosmos, it still must be in the cosmos because there is simply no other place to go to outside the cosmos. The friction that once produced the heat in the time before and during the Big Bang period still today actively participating as mass.

Mass is the result of let us call it "stationary friction" which is the relation between two cosmic objects having motion inequality but still sharing space within space. By creating friction through the bringing about of any form of motion discrepancy between objects in any such a test performed today such friction coming about will produce heat and the heat will result in space forming. In such contact between objects in different speeds that such motion discrepancy produces to cause destruction of matter in space and heat comes about. In that the net result eventually leaves space created when overheating material no longer fills the space after cooling sets in. The cracks showing in the cooled material afterwards is a result from the overheating of the material that created the extra space and then reset the occupation to what it was before. But the material that is reducing from retracting used space with the becoming colder again leaves cracks behind on the surface that proves that there was more space filled when the material was heated than what it had before the material was heated compared to the decrease in space after the material; again cooled down back to what it was before the material was heated. Evidence of this is evident in all supersonic aircraft as the fuselage forms cracks in the body structure of such an aircraft. The outcome of this heat is that when cooled the material occupies slightly more space than before. The grown space then tries to fit into the area it did fit into before the heating but with the extra space it employed when it was during in the heating process it shows afterwards as cracks. This takes us back to what Kepler said. In Kepler's formula it is the extending of the distance k that influences the time aspect T^2 which the supersonic aircraft does by its going supersonic and by shifting k from the previous location the Earth prescribed to the new k the aircraft implicate in according to the k that comes about since the distance in affect becomes longer. The aircraft produce a new time value T^2 in accordance with the Earth time factor T^2 because of the fact the Aircraft still shares space in space of the Earth with the Earth. The aircraft now has a bigger time vale T^2 in the space a^3 of the Earth using the Earth time so therefore the fuselage of the aircraft has to reduce its space a^3 it claims compensate fro the extending of k which it does by going faster as the extending of k will introduce a bigger time factor T^2 that will reduce the aircrafts occupying space since the Earths atmospheric space will not compromise and the aircraft still remains in the space of the Earth.

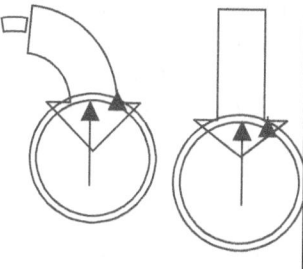

Liquid and motion

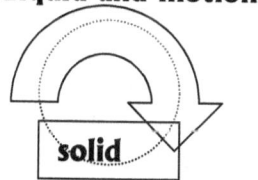

The bigger motion discrepancy there is between the Earth and any independent structure in motion within the control zone of the Earth will bring about a larger pushing of the secondary space the smaller object holds to slow down the motion the relevancy of the motion that the Earth in relation to the rebelling object which is captured by the space that the Earth controls. The object fights to retain independence while trying to while slow down the Earth as to either accelerate the motion of the rebelling object or to slow the Earth down. This is all an attempt by the Earth to increase the motion that the secondary object can apply or by friction reduce the object space-time to liquid heat. There is a specific border or a definite barrier where the motion differentiation becomes so critical that the incorrect transforming of motion can have the same effect on the incoming object as hitting a solid wall. Later on I shall show that it is in fact the equivalent of a solid wall the object will break if the object fails to enter through the small door there is in the area we call the atmosphere.

By having more heat per volume in ratio the material will claim and introduce new space that formed. Heat establishes space that expands. This truth science does not recognise. The claiming of more space and disposing of the space after cooling shows new space formed in the process of heat multiplying where there were no space before that which the material in the cool state afterwards cannot fill because of the void that came as a result of the material getting cold and contracting space, reducing the space as the space filled when the material was overheated. If material employs this as a basic technique today it was a basic technique back during the Big Bang. That evidence we can see in when material having a heat level amplifying upwards when motion difference brings on friction and such friction brings on heat. I do not share the view Mainstream science has that when matter and antimatter came into conflict the product that came from this just disappeared without a trace of any sorts. Two opposing issues came about, but both opposing issues are still present in the cosmos somewhere in a place where we are missing the presence thereof. Material is energy and energy is indestructible. However energy can change form...yes that we all know and energy may even hide appearances. Therefore we have to search for the new form in disguise. I believe the evidence is present at this moment in our Universe and I think I know where the evidence is. I believe I can show that it is a motion discrepancy that produced matter and anti matter and we do not have to go and look for non-exiting positrons and negi-protons (if I may be excused for using such bizarre terminology but it is fitting a bizarre statement which the first one is not my doing as my brain I did not make it). But a positron must produce a negative proton and such a performing sub atomic structure cannot be possible. By changing legions the proton must then perform gravity by rejecting material or if I am correct, producing space! I am about to prove that antimatter is in fact a process where the heat that became formed heat, which forms space, and therefore space has a valid substance other than being nothing. I go to lengths to make persons see that space cannot be nothing. This is a factor that science has to accept if Mainstream physics have the will to find solutions about the Big Bang. The motion between particles in a cramped space as the case was during the initial stages of the Big Bang would have brought on friction in space we cannot even calculate.

All the results coming from this we also find in nature. The extending of **k** is as much a contribution to gravity as he retracting of **k** is. It is the combination that forms gravity by motion. The result is that in the very beginning some matter particles produced gravity in their sustaining of independent singularity by applying motion which in some cases lead to the demise of some forming space-time by converting the where some compromised solidness. In order to install some form of coherency I shall for now and only for the moment call them the antiparticles that is then referring to those that became destroyed. Saying this I have to immediately reprimand myself in using such terminology because by doing that I am once again bringing in our human concepts of judging some or other form to our requirements thus placing me as apart of one side or part of another side, which I like or dislike (I think it is a normal human error but as I am the one preaching against it I should be the first to try and stop such human judging and picking sides). This route the one side took resulted in plasma forming on the one side and material on the other side. This was done because there was less control that confirmed the space and the volumetric space grew. Looking at the development from a less defining point of view may say that some got soft and others stayed firm. By having some softer than the other harder ones the softer one became a liquid. The notion or defining of a liquid is very relative because as solid as the Earth seems the Earth vibrates as a seemingly liquid during an earthquake. It forms waves many meters high just like it would be a liquid like the see. Afterwards when those in charge of damage control come to assess the damage it is hard to digest the destruction and damage because all liquid-likeness disappeared. The fact that the Earth had more motion in its own is the indication that during the earthquake the soil served as a liquid. This becomes a reality, which because of the mould-like ability a liquid has is mostly moving away or around the solid structure that remains apparent as static. In such a manner the more solid ones through the process known as electroplating could incorporate the softer ones. In moving from one position to another will commit the softer material to coming across as the liquid part. This electroplating motion is possible since electricity is gravity to some intense extreme. Electro motion or electro flow is the concentration of gravity to the limit where we will find gravity has the same intensity in the centre of the Earth as electricity has in the open. By removing material from the less dense and electroplating that which is removed from the less dense and then galvanise that softer material onto the harder material (which by the way is a very natural process taking place all around as a corrosion) the density of the liquid will demise in the liquid sector and the material will grow in the solid sector. I believe even to this day and throughout the rest of the Universe wherever there is space such space has to have motion and space cannot be what it is without having motion. With that in mind that is space-time. Space-time is space flowing on. Where there is motion in space, the motion through of all space is carried along by time in space. The plasma is transforming to material through the motion we named gravity.   By being electroplated onto material. By duplicating space in the process of establishing gravity the object does not reduce to a standard in occupying space that it had before the motion took place but by placing liquid heat into the form of solid matter the matter use the newly acquired heat through which to cool. By absorbers the liquefied material onto the solid material is thus freezing it into a solid to secure more material in the fight of combating overheating. In other words in the present time in our Universe gravity is freezing space to first become dense and form a liquid after which it then solidify the liquid heat by freezing the liquid into a solid state within the substance that is the atom. However, I do prefer to use heat as the term of choice and not plasma. Plasma is confusing since the variety of names in which it is identified. The process I just explained was the manner used by the cosmos as the cosmos came about and this is the manner that will repeat until such time as will the cosmos conclude its final motion. I believe that the first motion came about as singularity was without space and found irrepressible heat levels rising. By overheating it moved into space that was still non-existing and that had therefore produced motion to rebalance the heat. From this I also believe some material that came about from singularity overheating remained as particles forming atoms where there is this relation between the solid proton, the liquid neutron and the gas electron. This omission I clarify later on. After this phase where the atoms formed the compromises ensuring a successful passage in cosmic development, had come and gone another phase took its place, which had the liquid heat formed as it then presented a new innovation being unoccupied space. The development of space from liquid heat such fluid was becoming gas that is space with the ultimate gravitational relevance that space can carry. This was all contributing to the lack in contracting gravity promoting expanding gravity to those particles that applied lesser motion helping the extending of space to turn into heat that again turned into space. In this, of uncontrolled release of heat performing as softer space-time such release of space-time is the destroying of singularity secured in a unit, which again I believe (within reason) I do prove. I show that on the one side singularity introduce space-time, which confirms singularity and space-time makes contact with space-time not directly controlled by singularity or that, which is directly confirming singularity. This I conclude from studying Kepler's formula. I believe heat is the destructed form of material that overheated and this confirmation the atomic thermo explosions give us. But to realise that we must beforehand find what any and all space is and we have to accept that space is made of something.

When one applies heat to an object it expands. That is primary school science. This states that more heat applied leads to more space acquired by the heated object. In sharp contrast to this is the growth in space when heat levels rises but freezing brings about the opposite result. When I freeze an object that object reduces its occupied space as it shrinks. Removing heat reduces space. That comes directly as nature responds to heat and I can prove that easily. By expanding it accumulates space to increase the improving of the size of the material. The accumulating heat for the sake of securing singularity, accumulates the heat in the material whereas the freezing tarnishes the overheating symptoms by the removal of material in unoccupied space using external matter and setting motion to the material until it contracts into a form which we see as visible heat. The heat is in the form of dissolved singularity that became material as material used it as growth. That is why by freezing it will diminish the space as to accumulate the heat absorbing into the heat into the material to maintain the equilibrium needed in space. Taking this equation of nature to outer space we seem to confuse the natural law. With outer space as expanded as anything can get we regard outer space as incredibly cold. As heat sets in the normal flow will bring about expanding of heat into the form we think of as space that limits the heat overheating. Outer space is the very edge of expanding of space where heat cannot expand into space any more. Outer space is the limit, the epitome of expanding where heat meets space at the edge of all limits once more. Therefore being the representation of the very limit of expanding outer space has to be the hottest place there is. By applying heat to a kettle holding water, the adding of heat manifests as steam and steam is hot water that traded heat as it reviewed space. By allowing the receiving of the heat to continue the container will let loose steam in order to match the contributing of space. The manner in which heat expresses itself when confronted by overheating is to provide additional space through expanding of space. Outer space is outer space because outer space has expanded all it can it is still expanding to the speed of Hubble's $1/H_0$ which inevitably does not only affect far-off places where we cannot be, but effects us on a daily basis. As outer space is stretched to its limit, its limit will continue to stretch but while it is stretching it has to having more than it had before in that outer space holds the limit of heats expanding possibilities. Singularity has been expanding since way back when but that means singularity is still releasing heat as space-time that turns out as space in the universal time of outer space. In outer space heat cannot expand more therefore except for the continual growth that benefits all singularity throughout on a continuous bases concerning all outer space. If singularity expands when heated and there is a limit to the point it can heat, and that point of maximum expanding has been reached through the unleashing of heat, which is turning into space, we can with great confidence declare that space as the hottest space there is. Whatever expanding that is possible was done to secure the cooling and all cooling that can be introduced to bring about further cooling was performed that place is the hottest place there is just because of the shear implication that it can cool no further is as hot as it gets anywhere. If that is the case then it is safe to say that galactica then is freezing cold notwithstanding our concepts of heat and space and heat in space given to us by our collective culture and not by our ability to reason. It has expanded to the maximum that it can yet we think it is cold when it is the extreme there is in heat that introduced the maximum expanding. That is the contradiction of the century that much I do realise. At the inner core of a star all space shrinks into the oblivious but we consider it to be the hottest spot in the solar system. That just cannot be because when material shrinks it becomes cold and by shrinking into the oblivious it has to freeze into a gonner. Again that is the contradiction of the century. Why will that be? The space inside the star shrunk to the minimum there can be and that tells us the space has to be cold because of the shrinking took the space to a position where no space can shrink anymore. That shrinking of no more space can only be inside the inner star and in that region where gravity is at its strongest. With outer space as expanded as nature may allow the space that grew could only grow in conditions of heat because heat produces expanding and expanding is the result of heat coming about. Space shrink because it is cold: that we know and taking this law to the star centre it means regardless of our interpretation of hot and cold, that area in the star centre is as cold as it can get notwithstanding what our nature may tell us. Then obviously the same must apply to outer space for precisely the same reasons because it is so hot it can expand no more. We look at the hotness of space and the coldness of space but it is the relevancy to the solidity that forms the actual heat and cold limits. It is so hot no expansion can produce more space in outer space, as the outer space seems hot and quite the opposite reveals the true scenario inside the star in the centre of a star structure. That means the number of protons in motion has a lot to do with the cold and hot scenarios because where the protons are most dense the cold is in extreme. Only in the absence of space can so much heat gather in excess and the opposite is true about outer space where the least denseness found brings about the space in heat found in outer space. Our human selecting of hot and of cold and what is hot and what is not prevents us the clear vision we would have when truly understanding the applying temperature. Temperature comes about from spin and the smaller the spin density is the colder the space becomes because the more duplication produces the most cold. We think of outer space as 0^0 Kelvin but in fact it is as hot as no other place can be in the Universe. The coldest is where material is freezing solid as material does when frozen solid and the hottest is when by boiling the material is going into a gas with liquid being the intermediate position where heat acquires the space to perform as a flexible substance.

When we look at particles in outer space we see the particles being frozen. It is because there is such a severe contrast between the particles and the environment surrounding the particles and not the particles that is so frozen. The particles are in a gas state because the particles do not form a part that is part of the space unit. Hydrogen clouds of hundred of light years in diameter are a common sight in outer space. The heat we find filling space is not part of the space but like the particles the heat is a separate issue. That heat filling the space is another form of material that could conduce by diverting from space or marry the union of space by becoming more space. If it were that cold which we think it is, it would not have expanded into such a massive cloud but would have contracted forming a cube of frozen hydrogen. But as we can see the cloud expanded the gas as far as the gas can expand. That expanding is indicative of heat and has extremely little to do with gravity. If you are of the opinion that those hydrogen clouds will contract one day into forming a star, well then think again there is just no such a chance that that will ever happen because that is not gravity. Because outer space is completely overheating the condition it has in support of the particles makes the particles appear to be in a state of freezing but the particles is counteracting the heat limit it meets. However the particles did not contract the heat because the space in outer space contracted all the heat by means of expanding the heat into what singularity will appreciate. That is not because outer space is freezing the particles it is because in contrast to the heat of outer space the particles seems to be freezing.

The atom must be the utmost coldest and the proton is even much colder because when that cold escape it turns to heat forming space that no one can understand. When the spin of the atom allow the cold of the atom to release the heat it had it had frozen to space the atom holds but when this heat releases from the containing form of the atom it brings about much more heat than the Human mind can cope with. One may not look at the material and judge the surroundings. The fact that hydrogen remains a gas and so does helium in outer space must serve as enough proof that outer space is hot, regardless of our interpretation of the temperature gauge telling us what we wish to hear. One must look at outer space and judge outer space from the findings only considering outer space. If helium remains a gas it is hot. The removing of heat makes the centre of the Earth is cold although we see it as being terribly hot. The only reason why it can seem to be hot is because it is cold and in such a cold environment the heat can gather and space can collect heat because the particles find the surroundings extremely cold. The cold in the earth centre causes the concentration of heat by space reducing, as all cold surfaces tend to do. If it was hot the space within the Earth would expand and the space within the Earth where we think so much heat is concentrated does not expand therefore it must be cold. To gather and accumulate the space in a liquid means it became much colder being a liquid. Finding the surroundings terribly cold will allow the heat to gather and not expand but when the surroundings is hot it will not tolerate more concentration of heat and thus will expand to rid the balance of excess heat within space. Look at the sun and see how the sun turned the hydrogen to a freezing cold liquid at 6500 K. Hydrogen is in a fluid state within the sun and is colder than the hydrogen that is in a gas form in outer space. The sun is the coldest place in the solar system. That is when the protons oversupply the removing of space to produce the cold that is so apparent. By the reducing of space it can concentrate heat to a fluid state by producing the opposing cold that finally freezes the heat to a solid state. The expanding of space is a way of duplicating space without reducing space and by duplicating in the form of expanding it becomes just the opposite to duplicating by motion therefore reducing space by halving space in time. That is what gravity does. By motion space duplicates and by space halving it removes heat in space as well as by dismissing space. In all the applying of gravity space bites the dust. The density of the protons brings about space dense enough to harbour the heat in such quantities and visa versa applies in outer space.

We have to accept that the coldest place in the solar system is in the very centre of the sun because there the most number of protons sharing the least amount of space producing the coldest area that can allow therefore the hottest density of heat within the cold environment. It is the duty of scientist to look far beyond the ordinary and find why the inner star will be so cold and as to why outer space will be so hot while being seemingly so utterly cold. It is the duty of the professionals to find matters as they are and not as they would seem to look from a human vantage point. Later I will show why the star is so extremely cold and outer space is over boiling with heat expanding into more space. We have to see what forms space and why space can be the absolute basic container through which gravity can relay the influence it carries. We must come to realise that whatever forms space has to be that same ingredient which also is the basic component that forms the lot of everything in the entire Universe. When particles heat up the particles expand the space the particles hold to limit the heat rising. The particles claim more space when heated to preserve the cold. The claim to more space produces more space and reduces more heat. Such expanding brings about cooling. When particles heat or cool motion applies in some form. Motion started at a point when the Universe was extremely hot and there was no space. By introducing motion space formed and the lack thereof produced friction that became heat that became space.

The application of gravity that condenses space and bringing about heat by the compressing of space we apply in the way we go about tapping into the energy that nature provide. Internal and external engines combustion engines all rely on this application for harvesting motion by driving power. Compress space even today with a piston in a cylinder and then pump the compressed air into a container and such confining of space will increase the heat by the piston effort to reduce the space brought about in the container. The heat coming about inside the cylinder has no relevance to particles colliding because all compressor cylinders cool down colder because when that cold escape it turns to heat as the heat releases from space forming a secondary form of material forming space that no one can understand when the spin of the atom allow the cold of the atom to release into uncontrolled space. This release and unifying with space that heat does is the heat it had frozen heat because of the motion of spin to space that the atom holds remains in a frozen state under the guard of the spinning electron. But when this heat releases from the containing form of the atom frozen by the spin of the electron it brings about much more heat than the Human mind can cope with. One may not look at the material and judge the surroundings. The fact that hydrogen remains a gas and so does helium in outer space must serve as enough proof that outer space is hot, regardless of our interpretation of the temperature gauge telling us what we wish to hear. One must look at outer space and judge outer space from the findings only considering in the terms which outer space insists upon. If helium remains a gas it is hot. The removing of heat from the space that contained the heat makes the centre of the Earth cold. In our universe we see it as being terribly hot because the heat then forms a separate substance but remains a form of material (8) but that is because we see the heat and not the space derived from the separating of the heat. The only reason why the space can seem to be hot is because the space is cold and in such a cold environment the heat can gather in a much concentrated state and space can collect heat because the particles hold concentrated heat in the space separating the particles. By removing such high concentration of heat from the space that use to be expanded heat, the space then must contradict the heat by being extremely cold. We look at the heat in the space, which by that time is another form of material and find the surrounding heat in the space hot while the space is extremely cold. The cold in the Earth centre causes the concentration of heat by space reducing, as all cold surfaces tend to do. But the proton contributes that reducing of space. If it was hot the space within the Earth would expand and explode but the space within the Earth where we think so much heat is concentrated is so much it does not expand therefore it must be cold. To gather and accumulate the space in a liquid means it became much colder when the space parted from what then is being a liquid. Finding the surroundings terribly cold will allow the heat to gather and not expand but when the surroundings is hot it will not tolerate more concentration of heat and thus it will expand to rid the balance of excess heat within space. The concentration or release of space with heat or space from heat is a direct contribution of the singularity in control of the space-time. The regard of the singularity stipulates the conducing of heat in space or the release of heat to form space by means of bisecting the occupied space. Look at the sun and see how the sun turned the hydrogen it holds captured in its atmosphere to a freezing cold liquid at 6500 K. Hydrogen is in a fluid state within the sun and yet it is still colder than the hydrogen we find in outer space that is in a gas form in outer space. The sun is without any doubt the coldest place in the solar system. That is when the protons oversupply the removing of space to produce the cold that is so apparent in the heat levels that do not join the spell. By the reducing of space it can concentrate heat to a fluid state. By producing the opposing cold that finally freezes the heat to a solid state we find that is what matter is. The expanding of space is a way of duplicating space without reducing space and by duplicating in the form of expanding it becomes just the opposite to duplicating by motion therefore reducing space by halving space in time. That is what gravity does. By motion space duplicates and by space duplicating the material must be by dividing or bisecting - halving it removes heat in space as well as by dismissing space and in that concentrating heat. In all the applying of gravity space bites the dust. The density of the protons brings about space dense enough to harbour the heat in such quantities and visa versa applies in outer space.

We have to accept that the coldest place in the solar system is in the very centre of the sun because there the most number of protons sharing the least amount of space producing the coldest area that singularity can allow therefore bringing about the hottest density of heat within the cold environment. It is the duty of scientist to look far beyond the ordinary and find why the inner star will be so cold and as to why outer space will be so hot while being seemingly so utterly cold or hot in humanly applied standards. It is the duty of the professionals to find matters as they are and not as they would seem to look from a human vantage point.

Later I will show in much better detail why the star is so extremely cold and outer space is over boiling with heat expanding into more space. We have to see what forms space and why space can be the absolute basic container through which gravity can relay the influence that it carries. We must come to realise that whatever it takes to form space it has to contain something that is the same ingredient, which also is the basic component that forms the lot of everything else in the entire Universe. When particles heat up the particles expand the space the particles hold to limit the heat rising. The particles claim more space when heated to preserve the cold. The claim to more space produces more space and reduces more heat. Such expanding brings about cooling. When particles heat or cool motion applies in some form. Motion started at a point when the Universe was extremely hot and there was no space. By introducing motion space formed and the lack thereof produced friction that became heat that became space. It is natural and it is simple and above all it makes believable sense.

The application of gravity is that which condenses space by bringing about heat with the compressing of space. We apply the progress we have as a species in the way we go about by our skills to unveil ways we can tap into the energy that nature provide. Internal and external combustion engines all rely on this application for harvesting motion by driving power. Compress space even today with a piston in a cylinder and then pump the compressed air into a container and such confining of space will increase the heat by the piston effort to reduce the space brought about in the container. The heat coming about inside the cylinder has no relevance to particles colliding because all compressor cylinders cool down with time moving and not necessarily with the loss or release of particles. It is not only the discharging of air that will reduce the temperatures inside the container but the time flowing bringing motion about where the motion is not about particles escaping but heat escaping in the replacing of the heat density (not the density of the particles forming the material content within the container) but the space that compressed to heat will also bring about that the heat displaces through the container wall to the outside. This is bringing about equilibrium where heat will always flow from more dense areas to the lesser dense areas. This has no influence on the status of the particles on the inside of the cylinder but only concerns the density levels of the particles inside versus outside. After the pumping of air increased the heat in the cylinder which even can go to dangerous levels, the heat will reduce back to room temperature when further pumping seizes and that stops further air movement into the cylinder and such surging of pumping air is what brings about heat stabilizing.

Mainstream physics ignored the clear connection completely, notwithstanding it being so very obvious. There is this far in their recognising of principles in natural physics not one single reference made to prove their appreciation of this matter. They are bent on particle colliding. When particles collide such collision forms an atomic thermo release and that action we call an exploding atomic bomb. What principle this argument about particles colliding ignores is that all atoms use negative charged electron forming the atomic limit on the outside forming a definite border to the boundaries of all atoms and in both electrons from different atoms are being negative charged. In being negatively charged it means both will come out and totally reject the other. The closer they come the more violent the rejecting will be and such rejecting is the production of heat that will turn to space. The electrons repel other negative charged sub atomic structures, which the electrons are that form the outer borders of all atoms. With all electrons highly negatively charged (being as negatively charged as any possibility will allow to match the utter extreme) such electrons couldn't touch. If a train in Japan floats on a cushion of air because of equal charges lifting the train into the air, how much more will atoms repel other negative charged electrons considering how tiny they are?

The particles entering the cylinder bring with them an envelope wrapping the atoms in space that is there to distance atoms from one another. Such space formed as a result of and under the conditions prevailing outside the cylinder walls. The balance at first favours the forming of heat from the space coming in and being reduced in the containing size they are squeezed into is reducing the space from what it was on the outside. The space distribution inside then has changed considerably and reduced a great deal compared to conditions outside and with the decrease of the space distribution that space then becomes excess heat on the inside.

The electrons will disallow any contact directly between atoms. No force can be big enough to enforce such touching. It is because of that contact rejection electrons bring about that science has to use an overload of neutral neutrons putting them in the atom nucleus to fake a complying of charges that will eventually lead to atom touching each other but that is through enticing a neutral stance which is enticing a positive overload for a short while. When the touching of electrons does take place the event is called a thermo nuclear reaction where heat is released in unmatchable quantities and the atoms in reaction dissolves into a liquid heat. But other than producing an artificial balanced bomb the touching of electron will never take place since the repelling that they would provoke amongst one another.   The increase of heat by the distribution of particles in space connecting space and heat to particles is a separate issue that has nothing to do with contained particles colliding because why does it stop when pumping is seized. This ratio of heat reduction is time connected as much as it is motion dependent. Motion reduces space by expansion as much as time contributes to space distribution by allowing the flow of heat.

This means it is not the particles touching one another in the cylinder that is bringing on the heat levels that is rising. Neither is it the particles that will eventually bring about the explosion that will follow, if safety measures are exceeded and should the pumping continue regardless of the danger rising. When the pumping stops the heat immediately starts the reducing thereof. Most important is the realising that every atom constitutes of two parts. In fact the entire universe constitutes of the two parts I am about to mention. On the inside there is a circle that contains the sphere and holds material in contact with singularity. On the outside there is heat surrounding the inner material part within the sphere and distance the inner material from the space between it and the next atom. The electron forms the division between heat uncontained and heat contained. This is why the Roche factor is so very important. There can be friction between particles in reduced space under controlled circumstances where such particles are grouped together in a unit and as a unit elects a group singularity forming the centre of the chosen form of the unit. However, there can be no friction between particles of atomic dimensions as a result of what the cosmos produces to contain a solution to this problem during the Big Bang.

The Universe separated heat from material by covering the exterior of material with heat that forms space. Some material became softer by uncontrolled overheating while others remained more solid by containing form through controlling the overheating.   On the outside of all elements there are a layer that is the heat the element uses in relation to place relevancies between such an element and the rest of the cosmos. On the inside there is almost as much flexing available but we shall deal with that statement in a while. That which we think of as elements that are being a solid or a liquid or a gas is very much untrue since all elements are either solids or liquids or gas and none of them are a "natural" of any form mentioned above. It is a condition the element applies to secure a relative position under specific conditions. That is why spacecraft entering the atmosphere or that is by passing the Mach limit such machines get covered with a heat blanket. The space surrounding the craft becomes liquid as the space becomes more intense in concentrated space that forms heat. There can be no particle in friction and even more so way up there in the atmosphere at the altitude where the cosmos meets the atmosphere just because the particles up there are so sparsely distributed in that part of the atmosphere. Above and beyond this lies the fact that all the so called air particles are very volatile and excitable by nature and they are known to turn the slightest heat into rapid motion thus establishing a scene where the particle that supposedly are in contact with the aircraft sheeting will move away from the hot incoming aircraft. If then not for any other reason then it is because the particles are highly volatile and acceptingly sensitive to heat. Airborne particles are prone to motion just because it is the airborne element nature to change heat into motion and the motion comes about from their sensitivity to duplicate. No particle in the air being part of the space we call air which is in a free floating in that air can produce friction because of the volatile nature those elements have. Faced with the truth about this disinformation we have to search for other explanations that nature underwrites which forms a presentation more true to nature and will therefore be more sensible and less impractical. The craft's coming into the atmosphere produces a point where $a^3 = T^2k$ changes to $k^{-1} = T^2/a^3$ (the explanation is forthcoming a little later on) The distance separating the incoming object from the Earth centre reduces rapidly therefore the object start to descend towards the centre of the Earth. We must also acknowledge the fact that there is one specific point of specific entry where this will occur more than before. That point will rapidly increase the time factor where the incoming object crossed such a very visible border. By the reducing of distance k space a^3 will have to compromise in the relation of all the factors forming the equation since T^2 will very suddenly grow more acute. What happens is that the applying gravity reduces the space a^3 and the compromising factor comes about since the time factor T^2 moves back to a time where outer space was as dense back then as the density we now have within the atmosphere that then became as the Earth atmosphere. It is outer space that remained denser that what the outer space currently is. I am now referring to a process that I introduce as this letter unfolds which is by nature completely different to what is accepted by mainstream science. That which I refer too came about at a point just before the Earth established an atmosphere that grew through gravity and by the measure of the Earth gravity became separated from the atmosphere. While the gravity of the Earth contained the space surrounding the Earth in a much denser packed envelope the area not under the direct influence of the Earth governing singularity became more spacious. The contained Earth atmosphere grew denser as the solar system developed into what it is today. As the atmosphere released from what we think of as outer space that release from outer space made the atmosphere much denser and the space above the Earth which is using a reducing time factor and that makes the Earth more compact. That established the T^2 factor to be that more condensed when one compare in ratio the density with outer space. The density at the time there was when the separation came about in outer space at the time of such parting outer space allowed objects to move away. This parting brought a barrier that is in place between the Earth and the outer space and any object coming from outer space into the Earth's atmosphere. The incoming object then would have to reduce the measure of the space the craft holds as the containing singularity set new standards applying to the incoming object with which the craft then needs to affirms its form and its status within the contained space of the Earth. The reducing will then suddenly no longer use space as the compatible factor but the focus will shift to the time factor that dictates to the space what the space can be. Such reducing comes from the switch there is in space – time where it was in outer space performing as being $k = a^3 / T^2$ to what it has to be within the Earth atmosphere $k^{-1}=T^2/a^3$. When the atmosphere grew apart from the outer space there are two ways of looking at the event. One can think that outer space expanded by the implication of the Hubble constant or that gravity withdrew the atmospheric space of the Earth at the time that the parting of space came about. But however you look at it there was a time when both outer space and the Earth's atmosphere shared equal density as we find it still applies on the moon and on Pluto. Then the Earth became dynamic and now they do not share any density at all. Things were overall more compact back then than at the present time and that included all things in the Universe. The space component is reducing the time component by compacting space to alter the space – time ratio. This is portrayed by Kepler's formula $a^3 = T^2k$ It shows space as the density of space decreases. The Earth still compact space by reducing the volumetric confinement of space $T^{-2} = k / a^3$. This we call the atmosphere, as the atmosphere becomes denser towards the Soil of the Earth. There is a change in the time component. Most evident of this is when studying the pendulum. Just as we can see in the pendulum swinging, we can see that the swing reduces. Such reduction is because as the space diminishes every time the arm rocks from side to side. With this there is proof that in the developing atmospheric space of the Earth the ratios change from outer space. This is proved by the pendulum arms that Galileo's experiment used to show that the swinging pendulum indicates $k^{-1} = T^2/a^3$. Further more it proves that Galileo was correct after all and unnoticed by science Kepler helped Galileo prove Galileo's point. In this the net outcome establish Kepler as being correct and the Newtonian argument of friction brought on by gasses fall apart which is at that altitude where such friction supposedly should take place, the material in friction does not even present in the atmosphere. But science will stubbornly cling to the old theory with persistency that would warm any warring Field Commander's heart. Every element stands in different regard to the heat surrounding the material, which makes us consider the material to be either a gas or a liquid or a solid. The material in every element there is as such is all three forms and not of the forms one at all. It is the way under which the circumstances is presented that the element allows the heat to gather and accumulate as the surrounding heat occupying he surrounding space. Every particle is unique in the way it regards the heat to material ratio and how much heat it uses to form either the gas liquid or solid state. If space a^3 declines then so must motion in relevance will have to compensate by reducing k and limiting T^2 because space a^3 must always be equal to motion T^2k

Galileo proved space-time in that space diminish as space has to compromise as to sustain the flow of time but time slows down in stronger gravity.

$$a^3 = T^2k$$

$$\tfrac{1}{2}\,a^3 = T^2\tfrac{1}{2}\,k$$

With gravity applying space reduce as time duration increases and that was what Galileo's experiment proved. It proved $a^3 = T^2\,k$

As

$$\tfrac{1}{4}\,a^3 = T^2\tfrac{1}{4}\,k$$

the space surrounding the Earth, which we call atmosphere reduce in volume of space the

Galileo substantiated Kepler's findings that that space a^3 correlates directly to time T^2 when space a^3 compacts, with the decline of k reducing, the swinging arm of the pendulum that maintains time T^2. (1) The swinging arm will not, but reduces the space it moves through move slower when the distance k changes. (2) That proves that space and time $a^3 = T^2\,k$ is directly related.

heat content rises as much as that the space holds heat having the heat

rising by the same token. By becoming less the space also become hotter. The ratio there is between space and heat increases as space in measure reduces. We have to learn to see heat where the heat in the space has two different identifiable substances. We also must see material holding space to be different from the space holding the material. We must see material to be different from the heat covering the material and compromising the space that produces the format of material in being a solid, a liquid or a gas. This changes in the state of materials holds a direct relation to the heat that also claims a steak in that space.

On the outside of all material the density provides a distance in space vowing between objects. That density also introduces heat as part of the distance of the space that is in place under the specific conditions applying. This is density because in the cold particles will be closer and when hot (3) particles will be further apart. By performing motion through pumping air into a container the pump collects particles and rush the particles into the cylinder, which is just a cylindrical metal container. By (4) removing air from the atmosphere and squeezing that air into a container it leads to the reducing of the space between the particles *(5) in the cylinder when compared to particles outside the cylinder. (6) As the container fills the space (7) that was space meaning it was keeping particles away from each other at a certain distance turns to heat in a ratio to the square (8) as the compressing removes space and the material density within the space increases because the material density rising forms liquid heat. This further aggravates the heat brought in with the material in the pumping process because the compressing of space is adding to the presence of the heat by accumulating even more concentrated liquid heat..

To elements hot and cold as influencing substances are outside influences that do not apply to the core of the atom. The atom constitutes of densely frozen space flowing or liquid space and releasing(9) the liquid space into gaseous space (10) When insufficient control leads to uncontrolled expansion of such liquid heat. This is all singularity governed from all centres involved. Heat and space are influences outside the proton but we may imagine that with in the proton nucleus it is bitterly cold. The heat or space will surround the atom on the outside, but has clearly no influence on the inside of the atom and therefore of the star. The star is in every sense the atom forming the star. Atoms will reluctantly compromise by reducing space but this compromise in solid structures such as the atom is will totally depend on the singularity that rules on the applying conditions. The heat or space is a state that somehow extends beyond the electron, which does not influence the proton or change the proton. (1) Only gravity coming from massive numbers of protons working simultaneously can remove space from the inner atom. The atom cannot compress but will withstand the worst pressure there can be. Remember a star cannot have pressure. Neither is there any possibility of atoms touching. Only space in the form of gas surrounding individual atoms can to a certain measure compress.

The heat in concentration or the manifesting as space is neither hot nor cold because the proton presents eternal cold. Heat is an exterior influence bringing about influence between atoms within the star.

A gas will connecting other another. and gas space think 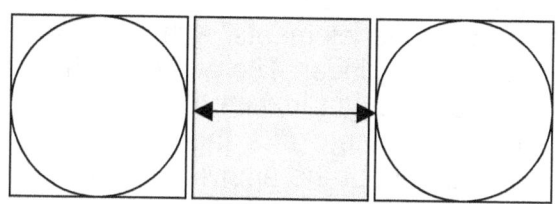 allow influences to charge as loosely bonding the positioning and forming objects occupying space next to one The gas will allow a lot of flexibility compromise because in the case of never becomes a premium. I should placing space at a premium would

apply when fusion comes into the picture. If a gas surrounds the hydrogen the gas will withstand as much compressing as can be induced by whatever force bringing about such compressing.

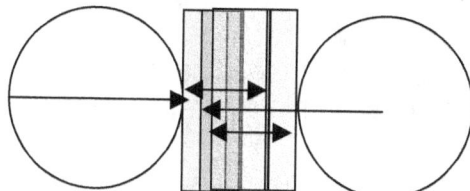

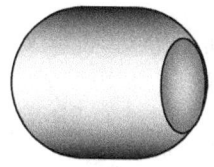 By pumping in air the molecules into an air-compressing container the molecules bring with every molecule entering a specific amount of heat from the outside that covered the molecule because all molecules entering is airborne molecules.

Scientists apparently in the know how of why what works express their opinion as one being that molecules run wild on the inside of the walls of the air container and bump each other while colliding and going mentally berserk in anguish and excitement. They maintain the argument that it is the clapping, fighting and general applauding of hostile, locked in particles that are in general behaving like British soccer hooligans that is going on inside the container. According to those Superior-Wise the general agony and anguish that the particles suffer while highlighting their complaints as they knock each other around ... well that is what is causing the friction, which is causing the heat to fill the container and fill the container to such an extent the heat that filled the container, is flowing out through the container walls. I wonder who thought this lot up? The person obviously was a tremendous thinker with true potent ability to dream an impossible dream with not a lick of logical argumentativeness mixed into any or in all of his thinking ability.

Every particle that is of the airborne type is of the airborne type because it favours an association with surrounding itself in an envelope of heat. That makes the airborne particles airborne particles. With that they disassociate with close proximity to others because they are airborne particles and airborne particles are what they are because they are cooking or boiling. Hydrogen, oxygen and nitrogen are airborne because they are in a worst state of overheating than water is when water is vaporised. The "gasses" are "gasses" because like "gasses" are "boiled vapour" even below 200 °C. Repressing the airborne elements as "gasses" is the same as conforming vapour in a cylinder. I witnessed once what happened when water vapour was accidentally contained as a vapour. The house is no longer classified as a built structure. When pumping the "gasses" the "gasses" brings in heat because in the vapour state they are highly associating with heat. When the airborne particles enter the cylinder they are no longer able to dissociate according to will, but forced to cluster while they will bring with their "boiling" "gas" status. This heat enveloping the elements is taking up space with particles that are enveloping them with heat as the compressor fills and that is what is compressing by compromising space where space. Space reduction becomes the norm needed to fill the container because the actual solidness of the airborne elements in own rite cannot compress in such a manner. That it places the burden of compensating to allow the increase in density on the space not occupied. The unoccupied space between the restraint particles deliver a gain in heat levels between all the particles and the next particle. Since all the pumping of air which is going into the cylinder the space soon gets cramped and compressing space forms heat. That is evident in all eternal combustion engines.

Since the heat is still under the specific conditions that was in place on the outside which was set by the atmosphere on the outside of the cylinder at the time when the pumping started the conditions in the air did not change by the pumping alone. As the pumping commenced the conditioning went on in the same manner, as it was the outside and taking those conditions inwards. The air takes the same heat with when the pump takes the air on the pumping journey. The pump is taking heat surrounding the elements with the elements and the conditions they were in, in the atmosphere into the cylinder. The pumping changed little about the heat applying under the conditions as it was brought in with the elements. In contrast to such conditions that were applying on the outside there arte then new rules that changed considerably. New dynamics on the inside brings about the applying of change to conditions. On the inside of the cylinder pumped with air forms new singularity elect that will enforce new controlling standards where all electrets are controlled by another set of laws. In the atmosphere where from the pump removed air the particles was widely spaced and the particles brought those conditions associated with a widely spaced atmosphere into where all are clustered in the inner confinement.

In the atmosphere there was different elements all bonded by a heat holding the space between the separated particles apart. These conditions was set by the Earth governing singularity ruling space-time at that point in the atmosphere. The ratio the particles were apart was precise and specific and preset by the sun burning heat into the air or not burning heat into the air at that specific point. But the heat was the fabric that connected the material because it connected to the material by a specific ratio that set distances we call air pressure. The heat between the particles were contracted to a certain degree by gravity and was not as utterly expanded as it would be in outer space nor as compressed as it will be in the container. Therefore some heat was in the space parting the particles in the conditions in the atmosphere at that very specific location. The Earth singularity brings about the gravity that determines the density of the "gluing heat" holding space between the particles in space. The solidness is the material is actually surrounded by heat that brings with the atom seclusion from other atoms by covering the atom and the surrounding space of the atom with the heat surrounding the material. It is the density of the heat in the density of the space that specifies the conditions defining every position every particle holds as space occupied relating to space unoccupied that is separating particles and filling space between the particles.

As the solid "stuff" is pumped in the solid "stuff' which is particles is covered by liquid "stuff" which is heat and together the solid and liquid "stuff" is pumped into the cylinder containing. This combination forms the composition we think of as pumped air. There is coming in as much heat with the pumped element material in the form of flexible space. That flexible space is that which is containing a specific degree of heat. As there is that much material coming in as a result of what the pumping is bringing in the ratio on the inside change somewhat. A lot of space turns to forming heat on the inside. The heat is in fact the outer part of the material coming along and is accompanying the particles as separate space but forms one unit. The particles in essence is solid when being without the heat containing the solid atoms and the atomic unit cannot compress as much as the particles cannot touch because all particles hold negative charged electrons to the outside. The electrons are all negatively charged. Excited negative electrons will not touch each other notwithstanding whatever any argument the wise in words wish to create arguments about confirming this idea. Negative particle subatomic or not will repel one another and the repelling will become more fears as the particles come closer. Unless scientists of high standings and other fiction writers can bring along antimatter and prove that the antimatter is hosting a positron, or what they seem to see as a positron there is no chance of electrons meeting. All electrons we no can only be what we then must call position because it is the only way that such a meeting can take place. Without the likes of a negatron and a positron meeting each other the two electrons sharing charges will not form a company. Without the likeness of such proof, (finding positrons next to negatron or what ever they will be named) the argument of touching atoms fall apart and will impress not even one clear minded thinking individual.

We have to recognise that with the space in the form spaces come, the space that is holding heat is part of the electron status. The heat in space is a part of the un-compromised space that is not yet fully extended to the maximum as space we find in outer space. Such space carrying pure heat within the space concentration also enters the cylinder chamber. The only flexing is the space containing heat as the space accompanies the molecules entering in a cover of heat. The space entering can enter as heat or as space since space and heat is the same thing. However in all circumstances there is a limit as there always is some limit to everything we find in nature.

The gathering of particles is also accumulating space, which then turns to heat by reducing the available space

When this comprising of space between particles should carry on within the compressed chamber there will come one point where the heat turns from space to liquid and then the liquid exceeds the limit that atomic bonding of the container can contain. What happens next in such a container that formed liquid heat puts us all in the centre of the star. There is only one exception and that is that where we are there is no control of a governing singularity that we will find in the centre of the star. The heat turning to liquid will exceed the atom bonding of the container walls. By exceeding the atomic gravity the heat between the particles overwhelm the bonding of the elements forming the cylinder wall, which takes the space as the atomic particles that forms the wall. It will supersede any limit the atomic gravity can bare in that bonding between all and every atom as they form a union that we see as the cylinder wall. The union that forms as the cylinder wall also become the union that forms the cylinder wall that holds the resisting imploding as well as exploding.

The wall is as strong as the particles forming the wall would permit and the atomic gravity will permit the union of the wall that formed to be. The collapse of such a wall under pressure that will come after the heat in space liquefied has to come about on the side of the material bonding together to form the wall. The wall then is the weakest spot because the liquid heat will never give way once the space turned to liquid. On the inside there is no governing singularity elect to contain the heat within the unit since the whole idea of a pressurised cylinder is artificial to the cosmos. The liquid in the space will cut any and all cylinder walls to shreds. It liquefies the space in the cylinder wall. The space forms heat and the heat forms liquid and the liquid cut the cylinder walls bringing about the explosion. If any reader is of the opinion that he or she has heard this before I advise such a person strongly to go back to Mainstream physics and find the official explanation about the material providing the pressure to bring about the explosion. Pressure has little or nothing to do with the whole procedure.

It comes down to density and this process is not the same or equal too the process that applies within the star. The star is filled with heat being in a liquid form and it is not filled with gas that is true but in the star there is a singularity elect that is strong enough to contain the liquid heat. With the likes of a strong enough governing singularity, the particles as such can become small enough to fit the whole Universe into the container because some time back the lot did fit into a space that was at the time the size of the container. While saying this we also must add that this may only be possible on the condition that we can find a container able o hold such contained heat. It is the release of the particles that concern cosmology since we accepted the Big Bang theory as the correct interpretation of facts, which of course it is.

What happens inside this cylinder is most important to cosmology because through the pumping of the air into the cylinder science consider conditions inside such a cylinder as the same as a star. Science holds the opinion that that cylinder is on its way too becoming a star. Such an idea is totally flawed. There is a big difference in the way the material inside the containers is contained. In the cylinder a big metal wall that is sometimes inches thick contains whatever is building up by means of pumping air from the outside to the containing inside. In contrast to this in the case of cosmic structures there is a massive pumping coming from a space less centre which does the pumping by contracting the space that prevent the escaping of whatever is contained. While the contained is contained there is a centre able to contain what the star wishes to contain. That is not the case with the air container. In the case of the star there is no requirement for the walls to contain with the any walls taxed by keeping it all on the inside. The inside of a star does not need a large metal jacket in order to contain what we find on the inside of the star. With that in mind it must be true that the Earth as all cosmic structures do, does not have pressure. For instance that which the Earth holds is contained from way within and there is very little chance of that which is contained has the opportunity to be escaping.

This we appreciate but also we know that whatever is inside the metal jacket cylinder container will eventually escape notwithstanding the best human efforts of containing. The escaping is just a matter of how long the escaping process will last. Clearly the two processes are so far apart in managed control; the two processes oppose each other totally. There is no comparison between the two. Even as far as the use of the term "pressure" is concerned. Therefore if we use the term pressure to indicate what the air container holds we are stuck with an idea how the system functions. This is totally contradicting what happens inside the cosmic structure. With the container not forming a visible border in the star there is containing coming from the very centre and this has the opposite of pressure within the limiting walls of the container in question. The air that came in to the cylinder as a gas will turn to liquid if serious pumping continuous and whenever the pumping exceeds safety limits the gas then turns to heat forming a liquid state. In the air container the density increases changing the cylinder on the inside from a gas to a fluid that becomes much more intense than the heat we think of as liquid heat produced by say an industrial cutting torch working with oxygen that is mixed with acetylene. The oxygen acetylene also forms a liquid flaming heat when the mixture is ignited but in the case of the cylinder the igniting will be spontaneous and burn through all the atomic particles.

The point of noteworthy is that at one very specific point in time the space becomes liquid and the heat returns back to space as it cuts through the cylinder wall bringing about "an explosion". The exploding is actually the space that heat returns to space and takes space from times when time was much closer to the Big Bang to what space is at the present. It is a Big Bang that is coming from then to that which is presently applying. The main issue to realise is that the pumping produces a density time factor that increases and that the density increases which turns the inside from gas to liquid. The cylinder moves back in time when the pumping starts and the moving aback in time realises another not containable Big Bang in the micro. It is not the oxygen or the hydrogen or whatever that fills the container that is a gas or a liquid that explodes, but it is the amount of space that turns to liquid heat that turns the container from a "planet" into a "star". Even the earth has already some flimsy liquid atmospheres in comparing to outer space. This is the only difference between planets and stars if you insist on having planets and having stars. The stage that the cylindrical container reach when space goes liquid is re-enacting the star with the difference that with the continuous pumping of air into the cylinder there is no sustaining or governing singularity and when uncontrolled the gas will turn to heat and the heat then all goes bang. By this measure mentioned the star is different because in the star such a liquefying is a long process that the star was awaiting for one eternity. That is the difference between compressor having a metal jacketed wall containing what is on the inside and having a pump on the outside that becomes most apparent in comparison with the star, In the case of the star the gravity within the centre of the star is where the star is having an inside pump. By the drawing or the gravity of the star, the star really shows its worth. This is a reaction coming about from the centre the star. Where space and motion ends is truly where the star starts to contain the inside without a possibility that the wall will burst as it goes bang.

The star does not allow escaping of material whereas the heat escapes from the container as it does from the star. When saying this we have to add that heart only escapes from young and incompetent stars such as the sun is. The star shows no relaxing in the process whereas in the event of no further increasing of pumping heat into the containing cylinder in any way possible and in the event that the pumping does not restart or continues again the process will rectify itself. Condition inside the container will once again return back to room temperature if pumping of air into the container stops. A balance will come about leaving the heat to escape through the container walls and then release as space outside the container. However, there is no loss of materials such as atoms that rushes out through the cylinder wall. This reversing of the flow of heat will escape through the wall relieving pressure without taking any of the confined particles along.

This is apparent because one can feel the heat coming through the walls as conditions inside the container walls strive to become equal once again to conditions as they are on the outside of the container wall. As the cooling goes on, the conditions on the inside of the container will again in time reach room temperatures inside the cylinder too to match the temperatures outside the cylinder. Once the pumping stops, the particles colliding and causing the friction then does not play a further role. They either become very calm unexplainably and without any apparent reason or the principle taught by Mainstream physics lacks truth. Never once could I ever find out what brought such calmness back to the material in the container without the particles having the benefit of a serious drag dos song that will be of a paralysing tranquil medication will bring on or having a real potent Sangoma (an African with or witchdoctor) present to enforce a calm. After all the science that teaches us that this heat increase comes from particles colliding and that is what is producing heat inside the air pressure container, the effort in getting them still and calm again after all the excitement must be a tricky operation. What is truly amazing about the becoming calm again is that the already over filled container does not continue the friction between the particles rubbing and brushing and through so many collisions stirring up more heat which was produced within the cylinder container as a result of what science teachings lead us to believe.

When the compressor is left by itself with no further filling or any releasing of the filling substance carrying on the temperatures inside the walls of the container stabilize as they go back to the same levels as what the temperature was on the outside. When all stabilizing comes about from time moving along and allowing the continuing of the stabilizing of the heat is under broken because there then is suddenly a motion of the air being released through the opening of a valve, such a sudden controlled releasing of the air under controlled conditions produces motion of the air as the air is relieved from the container through a release valve. The particles are unable to escape in the same process as the heat does. The heat flows through the cylinder walls and that means all the particles are still present and accounted for in the cylinder. When the particles are released through a releasing valve the release will create a flow of air, which starts the motion of air, which then will bring cooling. The pipes where through the air flows when released will cool to such an effect that pipes can and does freeze and block all airflow.

Two American Submarines were lost in this manner. It is not particles that have to take the blame but heat being released or admitted. If the cylinder is left undisturbed for a few days the stabilizing in the air within the cylinder walls will lead to re-establishing a total equilibrium and such establishing once again will take temperatures back to a freezing state as it cannot retreat heat that escaped space of the cylinder before the valve release came about. After the pumping subsides the space that is compromised then is allowing equilibrium to set in where the equilibrium equalises the particle density within the cylinder to that which is in the room temperature. The problem that results from this is that there are far more particles in the cylinder. That causes a much lower heat level or space in the cylinder than was the case when the heat was at room temperature. By stopping the flow and allowing an escape of heat the heat leaves space at levels where the levels in the cylinder is forced to be at a lower density then than what was applying on the outside. This flow of air in the releasing pipes show motion and such motion cools down space occupied. Motion is gravity and gravity is cooling of material into the oblivious. Through the motion coming from the releasing of air there is a much lower heat level in the cylinder. The motion of released air will force heat levels down when comparing the heat levels inside the container to what the comparing heat levels are outside the cylinder. The conditions will be directly in reverse as the space inside the container then is much colder by air motion than the conditions of the heat is on the outside. The natural flow of air from inside to the outside will change the conditions on the inside so much in spite of the air being overfilled on the inside and the comparing lack of density in the air on the outside.

If the container and pump is left without further human intervention or human influencing any circumstances developing such absence of interfering in the conditions as far as heat distributing goes will become and remain equal on either side of the cylinder wall. The compacting of air molecules is then still much higher on the inside than what the denseness is on the outside but this difference will not produce the same heat levels inside the cylinder than that what was achieved during and immediately after the pumping operations was performed. The temperature will only become affected when more motion contributes to changes in the balance. The releasing of air will extract heat from the process to a point where it will lead to freezing coming about in the narrow pipes where air is released and such airflow is the fastest.

The rebalancing will go the other way as was the case when the pumping was in place but when the released air is in natural motion the flow contracting heat flows from the outside to the inside of the pipes. But since the motion of airflow will be much faster as the air is released such motion contributes to cooling whereas the motion of pumping gathered heat. Although if science is correct then the heat could have started only when particles made contact and went bumping into one another with causing friction by their bumping and dancing. It is paramount for cosmologist not only to gather mathematical proof but also to apply such mathematical proof in amongst stringent natural laws and find not only what but also why certain events came about. The spin correlates time by supporting the flow of heat unoccupied to feed heat occupied. There is a definite correlation to establish balancing in space-time and one thing we know for sure is that if it applies today it is because it applied way back when it all started. There is a correlation between heat and space and heat in space, which is very far from being the same thing. Space contains heat because the atom is secluded including heat captured by individual singularity but space can also accommodate heat by producing expanding of space not captured directly by the control of singularity in the absorbing of heat. Heat will always flow from the hottest to the lowest region because the density of heat will bring about a flow such as water does in gravity. The mere fact that such flow of heat from hotter to a cooler region can take place makes it clear that heat will flow just like liquid does and find stabilizing as liquid does. Realising this must take our thinking back to a time when space was preciously little and heat was souring almost beyond control.

If we tackle an issue as unimportant as a container being full with air in the correct line of thought then it might enable us to draw direct parallels from this. Such parallels might enable us to see how fusion comes into the picture. If a gas surrounds the hydrogen the gas will withstand as much compressing as can be induced by whatever force bringing about such compressing. Gas can compress like nothing else can.

By heating an element the density will deteriorate releasing heat to form extra space and the decreasing density will bring about that more space becomes available within the confinement of the compressed area that is heating. But the more space does not come in the form of more material therefore the extra heat is extra space claimed by the same material occupying the space. If the confinement is such that it will not allow the growth of more space then the heat level will rise. By heating without allowing the liquid to form space then we find that liquid is even more uncompromising than material can be. The heat will grow by increasing heat levels in the liquid heat until a state will arrive where the liquid air will cut through any container that is man made. On the other hand it is true that gas will always compromise by giving away space to produce heat. With such knowledge we have to look at the functioning of stars once more but this time with a much more critical view. In fact having this realising it prompted me to look at pictures of the sun with much more critical gaze. What I saw was, the sun being a sloshing bowl of liquid flowing as fluid once more and then I realised from what I saw that the sun is liquid. If the sun is a liquid bowl then all other stars has a liquid inside! But we also know the density levels rises extensively as the sun's space decline towards the centre of the sun. That means the liquid will become more and more denser as the sun reduces space towards its most inner centre and the sun would have heat in a sub solid state down towards the centre. This changes every conception there ever was about the inner workings of stars. One can compress a gas to a state of liquid but then the gas is no longer a gas. The gas transformed from a gaseous state to that of being liquid. Gas can never compromise enough space to allow fusion to take place because there is far too much flexibility to bring about compromise, but liquids are quite another story. Before we come to that however, we must reclassify that which we think about when we think about a star as in this case the sun.

In the process there are relevancies. If outer space is - 276^0 C or 0^0 K then it can only be because other conditions apply elsewhere where the elsewhere will bring about that under the elsewhere conditions there is something else at 0^0 C and that can only be valid when another object is in an environment that is allowing that something in the other environment to reach 100^0 C. In other words there has to be a scale fitting all possibilities that goes from densely deep frozen hydrogen at -269^0 C to a boiling iron at $3000 ^0$ C. That is a spectrum but such a spectrum only has validity under very specific space-time applications. But if space stood alone and we gave outer space a temperature value of -276^0 C with no other references to compare such a statement has no validity because the number could be anything just as much as it means nothing. If there is no correlation between two factors that produces a range from zero to a maximum such a scale is as meaningful as much as it is senseless.

There has to be a relevancy to validate any figure in temperature used. If we pronounce anything being whatever it can only be valid if it is because being whatever requires not to be another whatever and therein then lies the value of being whatever. To bring about the relevancy is only valid when allowing us some comparing between the two of different and non-equals. With the one number standing alone such a number by itself has no meaning, not withstanding whatever connection there are between the two. With the sun being under another singularity in charge of different rules setting different standards to the markers we use on Earth the sun will have a totally different heat standard of freezing or boiling coming from the controlling singularity than we find applying in the governing singularity of the Earth.

The sun we have to admit is very different from the earth and Earth standards just will not apply where the sun is concerned. Making a statement that outer space was 10^{34} degrees in the shade during the Big Bang festival is rather meaningless because what was the other boundary then? There is a question arriving from this issue which is: Am I allowed to draw the comparing that the sun was $6500 ^0$ on the rim and $18 \times 10 ^9$ on the very inside at the time the Big Bang was 10^{34} in outer space? If that is the case and it has to be the case if science wishes to apply any meaning to 10^{34} in the shade at the time of the Big Bang, then the sun on the inside was an icc box at the time and during the Universe presenting the Big Bang presentation and on the inside the sun was one of the most potent freezers at the time. We then may regard the sun at the time of the Big Bang, as being a freezer storage facility, which was colder than the human imagination, will allow our perspective to go. And more compelling is the fact that this then makes more sense than anything anyone said about the Big Bang or that was ever previously said about the start of Creation.

The objective of a star or any sphere for that matter is compromising by compacting the inner space as one comes closer the centre. As the star is reducing the space it claims for individual use outside the wider Universe it is also condensing such space to the inside where such space claimed is providing a reference but what reference and to what will such a reference point refer? If the sun was a gas on the inside of the sun as science declare, then there was a lot of space going around within the confinement of the sun but that is also true that there then is not much filling the in between the solid substance also going around within the sun. I have seen where comparing was made between the sun and a tire. In such comparing I wondered what was in the mind of such a Master cosmologist when thinking of the star as a tire and a tire as the sun. If the sun is an air inflated tire what then is burning? The burning is the purist liquid of all, it is raw heat flowing like the liquid it is. That is the case where a liquid fills the space because the in between space is filled with a nicely flowing liquid. Liquids in the sun prove to be much more practical because liquid is more dense in certain ways that will solids can ever be. A solid will rapture when asked to compromise because after all that is what fusion is about. On the other hand will any a liquid just become more and become more liquid without giving way. Liquids did the compromising which lead to the Big Bang and after that there is no further compromising under pressure possible (if I am allowed this once to call the inside of a star pressure). That is why the strongest power engineers can use is hydraulic cylinders. By using a liquid within the star too the comparable gas outside the star and the solid being within the very inner star core, we can begin too see comparisons emerging. Remember that outer space gas is hotter than the inner space fluid of the star not withstanding our perception on culture formed by schooling and with memories that our culture is shouting to us in defying our wits because the truth is quite the opposite to what our minds are telling. When something gets hot space is added and inside the sun with all the filling of liquid and material space truly is an issue of scarcity. If it was not true that space is little in going around in the star then fusion could not have taken place.

Fusion is the product of space in the demise thereof and the demise of space brings about a need for space. That is the prelude in conditions applying upon which the Big Bang followed. But with such little space going around then the absence of space makes that the space which should be available, then has to be a priority and with space that little to go around in the sun, the sun must be bitterly cold. Again I stress that heat in abundance form space by the volume and that is one thing that is very absent in the sun. If space was in abundance fusion was not possible. Heat brings about space and looking into the sun we find that whatever the space there is, there is not plenty of that in the sun. With no space it means there is a cosmic cold raging in the sun. Do not for one minute think of size of the sun that we think is taking up space because that type of space is not anywhere in the cosmos. There is no big as much as there is no small and the space which we think of in terms of size and what size depends on is the precisely the incorrect human way of thinking because that is the human instinct to measure by size in terms of big or small. If we do measure by giving size in terms of big or small, hot or cold, near or far, and we then include our importance that we shower ourselves with as we put our being the centre of the Universe in the centre of the Universe. Then we set a standard judging from what we control but by putting us in a position in the Universe that makes our point irrelevant by which we use and whereby we judge the cosmos. If we make ourselves important in our eyes we become part of the cosmos and that makes us incapable, as we then are irrelevant about our judgemental concepts.

Having a position in the cosmos will translate to our incompetence as humans, which goes directly into our irrelevancy what we can use by which measure we judge. By placing me the human in the position that I, the human think I am in, then we give our position we think we have as forming the centre of the Universe such a relevancy by which we measure. By seeing us as life being the centre of the Universe we loose perception of what is valued in the Universe. That tendency we then have we must destroy to find meaning to our thinking about cosmology. We think of hydrogen as a gas in the atmosphere of the Earth and we transform that standard by which we think to the sun. If hydrogen is a gas here on Earth where we are then we are convinced that hydrogen is also a gas where the sun is and then the sun is a gas structure. How much more Biedermeier can we be when we are thinking in such terms…well atheism gives such thinking of the Biedermeier manner quite a go. If we continue to repeat that line of thinking we might as well move back to the cave because such backward thinking belong in the minds of cave dwellers.

The sun is as liquid as the sea is liquid with the difference that the sea uses water and the sun use pure liquid heat. That makes the sun a giant hydraulic pump and that removes the pretty weather system that we grant the sun from the sun. There is no winds blowing but there are rivers flowing at an astronomical pace. There then are no winds in the sun but there are rivers of flowing heat running and raging in the sun. By using hydraulic power within the sun instead of the presumed pneumatic gas as suggested by Mainstream science the rules on physics that is applying within the sun changes as much as day is different from night. In a hydraulic system the hydraulic power will only fail once the weakest spot in the solid breaks down. With a tough enough cylinder that will withstand all pressure pushing at it, something will give way and we know one hundred percent it will and cannot be the hydraulic fluid. The hydraulic oil will produce more fluid when overheating by acquiring more space from the heat asking for more space. However in such a case where we think of oil as a liquid the hydraulic oil is the solid substance, which we humans consider as being liquid and it is only performing as a liquid. It is not the liquid we should think of as being the substance in the sun. We also can see that the liquid that the oil is re-enacting or standing in on behalf of a true liquid. Being the liquid can only be valid when conditions apply to form and not to the state of the material. If one considers the qualities of liquid then we can ask what will happen when the liquid is pure and uncompromising heat in the purist form the cosmos can provide? What happens when that which is flowing is heat unable to break down?

In that case heat can only compromise by becoming more of what is containing the space and not less of the containing of space. By becoming more it also takes up more space and that is less uncompromising than what solids are. In a star such as Jupiter that has gas within the atmosphere, (let us call the atmosphere a gas this time because gas and liquid is a very grey issue) the pressure within the structure cannot apply a solid base to secure solids compromising space that will bring about fusion. The solidity that eventually brings about the collapse of occupied material holding space is unavoidable when confronted by an uncompromising liquid and by increasing the volumetric space the liquid holds it must come to a situation where something must break. In the case where fusion is in place everything is pushed to the extent where nothing can break. In this environment we know from the characteristics of hydraulics that the final collapse will not come from the hydraulic heat. The final collapse is that what happens when the solid atoms collapse from being two to being one structure and the solid singularity has to compromise independence. There are two in the space the one had previously and by fusion, the one gives way by denouncing its independence. This action can only present fusion when cold is at the limit there can be cold and the Universe at that point is as cold as it shall ever be and heat is at the very other end where heat can only peak as heat touches eternity at that very spot.

When outer space is 0^0 K then the sun is estimated to be 6500^0 K. That is the relevancy we bring in on a scale between the sun and outer space. On the one side the gas in outer space is -276^0 C or 0 K and then the sun changes the environment to what we know the sun is. The sun in our minds is the other side of any extreme where the sun fills space. But the material within is neither -276^0C nor is a frozen liquid on the atmospheric rim where outer space meets the suns atmosphere holding a freezing 6500 K. Deep within the sun the temperature seems to be18 X 10^6 K. It is only a relevancy with substance when humans relate the information to circumstances seen from the Earth. If standards applying to the sun were transformed to standards on Earth then no substance known to man including man in person would have any chance to exist because the sun harbour conditions that is beyond frightening as the conditions in the sun is hostile to life of any sort. Actually when thinking about it in a scientific manner this statement includes all conditions applying anywhere except on Earth and then not even all of Earth. By placing life anywhere and at a penny a dozen that notion put the manner science consider life and their life's ability to judge into question.

When any person and in particular a scientist is assuming a role as being a judge of conditions in the cosmos such judging can only be introduced when the person in judging divorced him or her completely from they're associated with life. We have to disregard whatever we feel about the importance of life and only pursuit with what importance the cosmos regard life. We have to eagerly pursuit the sun as our nearest star after we dislodged all mind concepts with life attributing to our way of thinking. This is not just necessary but it is a necessity. Where Mainstream science so desperately wishes to find life all over the Universe they should realise that even on Earth conditions differentiate where certain life can be and where other forms of life cannot be. Even the Earth is at some places hostile to all forms we think of as life. Life cannot be either in outer space or on the rim of the sun or in the very inside of the sun. When investigating that which forms the cosmic spectacular one thought that must never leave our minds is that of whatever is out there, we have this tiny spec and that is all. With that realisation we have to come to terms by excluding life when judging the Universe as well as all conditions we may be familiar with because such conditions will fit in and will swing our judgement when such thinking is suiting life. With standards applying in the sun however there must be one point that is zero to give the sun an accepted range. Say the sun would be 0^0C at the point we give it the norm of 6500^0 as that will be the lowest temperature that can be found within the confined space in the boundaries of sun. This will be because the sun is a secluded, isolated Universe detached from all other universes except the immeasurable universes we think of as atoms that group together and form the single universe we call the sun.

The sun holds all that is inside the universe that is the sun. From the singularity in charge of the sun there is one universe .,. that is the sun and whatever else is there, that which is there, is there to feed the universe of the sun. Under such conditions water that is so absolutely crucial in the mind of human's would be an unknown substance holding an unknown quantity as a reference. Outer space must then be minus 6776 K if the sun whish to use the scale we apply. But it does not work in that manner either because most of the substance that is providing life the support life depends on and life therefore have no claim to space-time within the sun. All readers might be agitated by my comment about life not being able to survive in the sun, but go back and investigate what rules scientist use when we investigate the sun. All rite I do admit that I am exaggerating about life being part of the sun but they (them in science) put seas and winds and even fair weather in the sun. We find winds blowing, gas clouds and such things. All conditions outside the Earth atmosphere are completely hostile and dangerously destructive to life. If we wish to obtain a clear mindset in our discovery of the cosmos, we should not try at first to discover other life wherever we wish to think we can find life in the cosmos before we discover the cosmos.

We have to set standards in regard of very Universe by setting a standard by which other following our thinking afterwards can truly apply our thoughts. The standards set can only match when all conditions are met with the criteria that that any specific Universe dictates to allowing life. We must then look for life by exactly the very same standards applying on Earth where we find life. We have to find a dot that holds water in all three conditions and we must be able to see the structure with our eyes. If we cannot see the structure those scientists then cannot boast that they are only using truths as science. They must then admit they are in fairyland and introduce their standings as such. We know water boils at a much lower temperature 100 km up in the atmosphere than it boils on the surface of the Earth and down below sea level it boils even at a higher temperature. This again supports yet again my argument about outer space being much hotter than anywhere else in the Universe because it needs less heat to apply to get water to vapour. We take water as the absolute element of prominence to our survival therefore in all the scales we apply we use water as the centre value and then in one single step further we think about it as being pivotal requirement for cosmic scaling not deliberately but just because we forgot to switch off out human connection. We focus on water as the standard applying whereas water in not found in any other structure. We even go as far as taking water to boil at 0^0 C because our thinking that way suits us in our human requirements but even that is a total abolishing of the truth.

Water boils where water is located and according to the density applying at such a point where that density in space-time will have a heat to space ratio that determines the boiling point of water. True as it may be that we have little use for water boiling at 5^0 C it does nevertheless boil somewhere at altitude at 5^0 C and some places higher up at far less even less than five degrees C. Putting water and our totally artificial manufactured perception of the boiling point thereof to where it is in a centre stage and we project that centre stage we think about manufacturing it even further to fit the entire cosmos underlines the desperate inadequacy of our reasoning about cosmic matters. We should not see how life might find the use for the boiling water but we must focus on the rules bringing about the boiling of the water. If water takes less time or heat to boil then water must be closer to boiling point in the highest atmosphere and that can only be if water up there is hotter from the beginning before the boiling process started.

The reason why water boils at so low a temperature at any high altitude must be from the more heat present in the space where the boiling is taking place. With more heat in less space less heat is required to heat more space to get the water boiling. I am referring to the actual space, which heat and material occupy, and is apart from the occupant heat, which is in a fight with material for space to occupy separately. I am referring to heat expanding to the point where the heat occupies all of the space and the space has more room to expand therefore the space took in all the heat it can use to bring about expand ing. If ever life will find a way to go down into the inner earth say at a distance of 1000 km we will find that at that point water only boils at many hundred of degrees Celsius notwithstanding the fact that the average temperature will also be many hundreds of degrees Celsius everywhere around. Conditions on Earth have such a variation, yet science has this tendency to standardise everything in the cosmos by applying constants that will fit the conditions we find applying in downtown New York on a pleasant sunny day in mid spring. If water boils that quickly the closer we get to outer space it should be a big indicator that water is naturally much hotter from the start in outer space. When the heat surrounding the outside of the molecule increases the heat inside the element has to reduce because there is a relevance attaching to the two opposing limits without the one opponent having any precise limit to show. As the outside fluid heats up, there is no breaking down of the substance because the fluid is the purist fluid there can possibly be. It is pure liquid heat. The heat will be more solid than the solid elements can be because the heat cannot give way any more than it already did at the event of forming conditions that realised the Big Bang. Just before the Big bang arrived it took all the heat occupied or not to form heat that would be able to bring about the forming of space as a compromise to the heat that was bound to destroy the Universe. By sustaining the conditions applying even before the Big bang there is not enough gravity left to produce more compromising from heat in a liquid form.

By heating the fluid the space the fluid holds becomes more and by heating the fluid the element becomes colder reducing the space the element holds. Space reducing is synonymous with becoming colder and becoming colder is about compromising space occupied. On the earth material will reduce space occupied that much and no more because the relevancy establishing the edges can only push the reducing that far and no more. But in the sun the conditions applying is a lot different and the relevancies can push that much further. As the elements enters conditions suitable to allow fusion one of the two factors surrounding the fusion will insist on all compromising of all space that the elements may have in separating between them or in holding the structure of the atom and in that act it supply the biggest compromise in the relevancy between the shrinking of material as cold or the space as liquid that removes space by heating more and that is to give space over to the fluid side that removes space from the material side. In receiving the space the heat sacrificed heat for space by acquiring a sudden space.

By acquiring the space it receives a cooling that will translate to dark spots on the surface of the sun once the heat again surfaces to the top. The dark is only contrasting the light because the dark allows less density therefore has more space. There is no actual hot or cold to be found anywhere in the cosmos. In sacrificing heat the liquid obtained space and by sacrificing space the elements in fusion reached the ultimate freezing temperature it could ever achieve. The elements never became hot or never frozen but the relevancies between solids and liquids. This is because mass has no role to play in the fusion process. Changing brought about condition allowing the heating and cooling to take place without ever taking place.

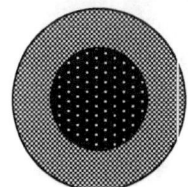

The solid	The Liquid	The gas
Hydrogen 1	melts at -259^0 C,	boils at -252^0 C,
Helium 2	melts at -269^0 C	boils at $-268,9^0$ C
LITHIUM 3	melts 180^0 C	boils at 1300^0
BERYLLIUM 4	melts at 1287^0C	boils at 2770^0C
BORON 5	melts at 2030^0 C	boils 2550^0 C

It clearly shows that there are groups of elements with much different relations to heat than other groups have. This is such a dominant part of Creation I truly am surprised our distinguished Academics never investigated the evidence in hand of elements forming different conditions that obviously totally devastates Newtonian claims on mass and weight producing gravity.

It is believed that mass produces gravity but according to my reckoning it is gravity that produces mass and mass is only the result of gravity. The less space a mass holds within the stronger the mass is subjected too the more mass there is in less space it to holds. Mass increases when the mass is surrounded by atmospheric heat where the heat will increase the mass as the additional heat will add onto the mass and as the heat will influence mass. By adding heat the object becomes more massive but also more spacious. In such an event the adding of material albeit heat increases the mass. But where gravity increases the mass increase as the space reduce. That shows a complete different tendency. Therefore mass does not produce gravity as science indicate by their formulas used in calculating gravity because gravity does not increase when heat influences mass. Heat stored in motion produces gravity. Any one not in agreement convince yourself by comparing the neutron star with the massive read giant and by you're acquiring a logic conclusion that is not tainted by your opinion about big and small you will be convinced of my correctness. Science go about in order to measure by calculation a Black hole they go and throw C^2 next to the dividing radius and throw the square onto the C by removing the square from the position the r normally have and they're doing that should then presents the speed of light acting as the retaining diameter. What the hell the speed of light, which is pure motion has to do with a diameter, which is merely a measure of a space distance and to top that their reconciling of the two is beyond my ability to substitute fact for imagination. With that they can the evenings fun and games because then they can cheat enough to make nonsense of the whole lot they play with.

Then they sit back and feel smart in the way they manage to cheat once more to prove their incorrect views correct because after all who will ever fly down a Black hole and return to support or deny their calculations. The gravity applied by the Black hole is a speed, measure comparing space in ratio to time taken because all gravity is speed. Then the speed that light has is gravity. The gravity of the light can be gravity as much as it at that very same time can be antigravity. What the hell has C^2 got to do with a Black hole because you can pop what ever nuclear device far away from a Black hole and it would be at the most and at the worst very much insignificant. The light will not even escape form the gravity of the Black hole but that has nothing to do with the diameter of the Black hole except on the condition that they agree that gravity is the reduction of space and that brings about a link. But that they never even suggests. When this became apparent that the radius of stars reduces as the stars develop through progress, someone was supposed to say: hey there is a dead rat I smell. For my saying so I am the clown in the courtyard, and the Academics see me as the one with the two dead brains cells and have no more to use as spare.